HANDBOOK OF OIL INDUSTRY TERMS AND PHRASES
2nd Edition

Handbook of
Oil Industry
Terms and
Phrases

By R. D. Langenkamp

 Second Edition

BOOKS Division of The Petroleum Publishing Company

Tulsa ● 1977

© Copyright 1974, 1977 The Petroleum Publishing Company
Box 1260, Tulsa Oklahoma 74101

Library of Congress Catalog Card Number 76–53234

ISBN 0–87814–032–8

Printed in the United States of America.
2 3 4 5 — 81 80 79 78 77

Preface to the Second Edition

The second edition of the Handbook was written for the express purpose of keeping abreast of the progress of an industry, international in character and global in its scope of operations.

Today, oil men and women are at work on several frontiers each as demanding in effort and innovation as any the industry has confronted in its 118-year history. Awesome, indeed, is the physical frontier of deep-sea, arctic, and jungle exploration. The scientific and technological frontier is demanding continued progress in synthetic fuels, and more imaginative techniques to recover larger amounts of oil from known reservoirs. And equally formidable is the relatively new frontier of political challenge by governments at all levels.

To note this unprecedented activity, the Handbook has been revised and considerably enlarged. The second edition has succeeded in drilling deep into the rich, saturated sands of the industry's lexicon and has come out of the hole with much that is new in technology, processing, equipment, and operating methods.

Acknowledgements

I can not deny myself the pleasure of publicly thanking David L. Dobie, second-generation oil man whose father Leslie Dobie came to Oklahoma before it was oil country and became a wild-catter and independent oil operator. David, also an independent, not only gave valuable criticism, making corrections and needed changes, but also added terms that otherwise would not have found their way into the glossary, leaving the work all the poorer.

Also, I must note in print what I have often expressed privately, an appreciation for the help and encouragement from Mimi, without whose unflagging interest the project might never have been flanged up.

A

AAODC
American Association of Oilwell Drilling Contractors

AAPG
American Association of Petroleum Geologists

ABANDONED OIL
Oil permitted to escape from storage tanks or pipeline by an operator. If the operator makes no effort to recover the oil, the land owner on whose property the oil has run may trap the oil for his own use.

ABANDONED WELL
A well no longer in use; a dry hole that, in most states, must be properly plugged.

ABSOLUTE ALCOHOL
One hundred percent ethyl alcohol

ABSORPTION OIL
An oil with a high affinity for light hydrocarbons but containing few if any of the light compounds composing gasoline or natural gas. The oil used in an absorption plant (q.v.).

ABSORPTION PLANT
An oil field facility that removes liquid hydrocarbons from natural gas, especially casinghead gas. The gas is run through oil of a proper character which absorbs the liquid components of the gas. The liquids are then recovered from the oil by distillation.

ABSORPTION TOWER
A tower or column in which contact is made between a rising gas and a falling liquid so that part of the gas is taken up or absorbed by the liquid.

ACCELERATED AGING TEST
A procedure whereby an oil product may be subjected to intensified but controlled conditions of heat, pressure, radiation, or other variables to produce, in a short time, the effects of long-time storage or use under normal conditions.

ACCUMULATOR
A small tank or vessel to hold air or liquid under pressure for use in a

hydraulic or air-actuated system. Accumulators, in effect, store a source of pressure for use at a regulated rate in mechanisms or equipment in a plant or in drilling or production operations.

ACETONE
A flammable, liquid compound used widely in industry as a solvent for many organic substances.

ACETYLENE
A colorless, highly flammable gas with a sweetish odor; used with oxygen in oxyacetylene welding. It is produced synthetically by incomplete combustion of coal gas and also by the action of water on calcium carbide (CaC_2). Also can be made from natural gas.

ACID BOTTLE INCLINATOR
A device used in a well to determine the degree of deviation from the vertical of the well bore. The acid is used to etch a horizontal line on the container and from the angle the line makes with the wall of the container, the angle of the well's course can be arrived at.

ACID SLUDGE
The residue left after treating petroleum oil with sulfuric acid for the removal of impurities. The sludge is a black, viscous substance containing the spent acid and the impurities which the acid has removed from the oil.

ACID TREATMENT
A refining process in which unfinished petroleum products such as gasoline, kerosine, diesel fuels, and lubricating stocks are treated with sulfuric acid to improve color, odor, and other properties.

ACID-RECOVERY PLANT
An auxiliary facility at some refineries where sludge acid is separated into acid oil, tar, and weak sulfuric acid. The sulfuric acid is then reconcentrated.

ACIDIZING A WELL
A technique for increasing the flow of oil from a well. Hydrochloric acid is pumped into the well under high pressure to reopen and enlarge the pores in the oil-bearing limestone formations.

ACOUSTIC PLENUM
A sound-proof room; an office or "sanctuary" aboard an offshore drilling platform protected from the noise of drilling engines and pipe handling.

ACOUSTIC REENTRY
A method used in deep-water operations offshore to reposition a drill ship over a bore hole previously drilled and cased. The technique employs acoustic signals to locate the pipe and guide the ship into position.

ACRE-FOOT OF SAND
A unit of measurement applied to petroleum reserves; an acre of producing formation one foot thick.

ACS
American Chemical Society

ACT SYSTEM
Automatic Custody Transfer System (q.v.)

ACTUATOR
See Operator

ADA MUD
A material which may be added to drilling mud to condition it in order to obtain satisfactory core samples.

ADAPTER
A device to provide a connection between two dissimilar parts, or between similar parts of different sizes. *See* Swage.

ADDITIVE
A chemical added to oil, gasoline, or other products to enhance certain characteristics or to give them other desirable properties.

ADSORPTION
The attraction exhibited by the surface of a solid for a liquid or a gas when they are in contact.

ADVANCE PAYMENT AGREEMENT
A transaction in which one operator advances a sum of money or credit to another operator to assist in developing an oil or gas field. The agreement provides an option to the "lender" to buy a portion or all of the production resulting from the development work.

AEC
Atomic Energy Commission

AERIFY
To change into a gaseous form; to infuse with or force air into; gasify.

AFRA
Average freight rate assessment (for tankers).

A-FRAME
A two-legged, metal or wooden support in the form of the letter "A" for hoisting or exerting a vertical pull with block and tackle or winch line attached to the apex of the A-frame.

AGA
American Gas Association

3

AIChE
American Institute of Chemical Engineers

AIMME
American Institute of Mining and Metallurgical Engineers

AIR-BALANCED PUMPING UNIT
An oil well pumping jack equipped with a piston and rod that works in an air chamber to balance the weight of the string of sucker rods. The device is attached to the well end of the walking beam and, acting as a shock absorber, does away with the need for counterweights on the rear end of the walking beam.

AIR BOTTLE
A cylinder of oxygen for oxyacetylene welding; an air chamber (q.v.).

AIR CHAMBER
A small tank or "bottle" connected to a reciprocating pump's discharge chamber or line to absorb and dampen the surges in pressure from the rhythmic pumping action. Air chambers are charged with sufficient air pressure to provide an air-cushion that minimizes the pounding and vibration associated with the pumping of fluids with plunger pumps.

AIR-CUSHION TRANSPORT
A vehicle employing the hovercraft principal of down-thrusting air-stream support, developed to transport equipment and supplies in the arctic regions. The air-cushion protects the tundra from being cut by the wheels or treads of conventional vehicles.

AIRED UP
Refers to a condition in a plunger pump when the suction chamber is full of air or gas blocking the intake of oil into the chamber. Before the pump will operate efficiently, the air must be bled off, vented to the atmosphere through a bleeder line or by loosening the suction valve covers to permit the escape of the air.

AIR INJECTION METHOD
A type of secondary recovery to increase production by forcing the oil from the reservoir into the well bore. Because of the dangers inherent in the use of air, this method is not a common practice except in areas where there is insufficient gas for repressuring.

ALIPHATICS
One of the two classes of organic petrochemicals; the other is the aromatics (q.v.). The most important aliphatics are the gases, ethylene, butylene, acetylene, and propylene.

ALKYLATION PROCESS
The process of making gasoline-range liquids from refinery gases, i.e.,

isobutane, butylenes, and others. The resulting alkylates are highly desirable components for blending in premium grade gasolines.

ALLOWABLE
The amount of oil or gas a well or a leasehold is permitted to produce under proration orders of a state regulatory body.

ALL-THREAD NIPPLE
A short piece of small-diameter pipe with threads over its entire length; a close nipple.

ALUMINUM CHLORIDE
A chemical used as a catalytic agent in oil refining and for the removal of odor and color from cracked gasoline.

AMERIPOL
The trade name for products made from a type of synthetic rubber.

AMMONIUM SULFATE
A salt having commercial value which is obtained in the distillation of shale oils.

AMYL HYDRIDE
This fraction in the distillation of petroleum was used as an anesthetic by J. Bigelow and B. Richardson in the year 1865.

ANNULAR SPACE
The space between the well's casing and the wall of the bore hole.

ANNULUS OF A WELL
The space between the surface casing and the producing or well-bore casing.

ANODE
A block of non-ferrous metal buried near a pipeline, storage tank, or other facility and connected to the structure to be protected. The anode sets up a weak electric current that flows to the structure thus reversing the flow of current that is associated with the corrosion of iron and steel. See Rectifier bed.

ANTICLINAL FOLD
A subsurface formation resembling an anticline.

ANTICLINE
A subsurface geological structure in the form of a sine curve or an elongated dome. The formation is favorable to the accumulation of oil and/or gas.

ANTI-KNOCK COMPOUNDS
Certain chemicals that are added to automotive gasolines to improve

their performance—to reduce "ping" or knock—in high-compression internal combustion engines. Tetraethyl lead is one well-known compound.

API
(1) The American Petroleum Institute; (2) The proper way to do a job; "strictly API."

API GRAVITY
Gravity (weight per unit of volume) of crude oil or other liquid hydrocarbon as measured by a system recommended by the API. API gravity bears a relationship to true specific gravity but is more convenient to work with than the decimal fractions which would result if petroleum were expressed in specific gravity.

APPRAISAL DRILLING
Wells drilled in the vicinity of a discovery or wildcat well in order to evaluate the extent and the importance of the find.

APRON RING
The bottom-most ring of steel plates in the wall of an upright cylindrical tank.

AQUAGEL
A specially prepared bentonite (clay) widely used as a conditioning material in drilling mud.

AQUIFER
Water-bearing rock strata. In a water-drive oil field, the aquifer is the water zone of the reservoir.

ARC WELDER
(1) An electric welding unit consisting of an engine and D.C. generator; usually skid mounted. (2) A person who uses such a machine in making welds.

AREAL GEOLOGY
The branch of geology that pertains to the distribution, position, and form of the areas of the earth's surface occupied by different types of rocks or geologic formations; also, the making of maps of such areas.

AREOMETER
An instrument for measuring the specific gravity of liquids; a hydrometer (q.v.).

ARGON
An inert, colorless, odorless gaseous element sometimes and in some locations produced with natural gas.

AROMATICS
A group of hydrocarbon fractions that form the basis of most organic chemicals so far synthesized. The name aromatics derives from their rather pleasant odor. The unique ring structure of their carbon atoms makes it possible to transform aromatics into an almost endless number of chemicals. Benzene, toluene, and xylene are the principal aromatics and are commonly referred to as the BTX group (q.v.).

ASK SYSTEM
Automatic station-keeping system; the name applied to a sophisticated drill-ship positioning technique consisting of subsurface acoustical equipment linked to shipboard computers that control ship's thrusters. The thrusters fore and aft reposition the ship, compensating for drift, wind drag, current, and wave action. See Dynamic stationing.

ASME
American Society of Mechanical Engineers

ASPHALT
A solid hydrocarbon found as a natural deposit. Crude oil of high asphaltic content when subjected to distillation to remove the lighter fractions such as naphtha and kerosine, leave asphalt as a residue. Asphalt is dark brown or black in color and at normal temperatures is a solid. See Brea.

ASPHALT-BASE CRUDE
Crude oil containing very little paraffin wax and a residue primarily asphaltic. Sulfur, oxygen, and nitrogen are often relatively high. This type crude is particularly suitable for making high-quality gasoline, lubricating oil, and asphalt. See Paraffin-base crude.

ASPHALTIC PETROLEUM
Petroleum which contains sufficient amounts of asphalt in solution to make recovery commercially practical by merely distilling off the solvent oils.

ASPHALTIC SANDS
Natural mixtures of asphalts with varying proportions of loose sand. The quantity of bituminous cementing material extracted from the sand may run as high as 12 percent. This bitumen is composed of soft asphalt.

ASSEMBLY
A term to describe a number of special pieces of equipment fitted together to perform a particular function; e.g., a drill assembly may include other pieces of down-hole equipment besides the drill bit, such as drill collars, damping subs, stabilizers, etc.

ASSIGNEE
A recipient of an interest in property or a contract; in oil and gas usage, the recipient of an oil or gas lease; a transferee.

ASSOCIATED GAS
Gas that occurs with oil, either as free gas or in solution. Gas occurring alone in a reservoir is unassociated gas.

ASTM
American Society For Testing Materials

ASTM DISTILLATION
A test of an oil's distillation properties standardized by the American Society For Testing Materials. A sample of oil is heated in a flask; the vapors pass through a tube where they are cooled and condensed; the liquid is collected in a graduate. When the first drop of "distillate" is obtained the temperature at which this occurs is the "initial boiling point" of the oil. The test is continued until all distillable fractions have distilled over and have been measured and their properties examined.

ATMOSPHERE, ONE
The pressure of the ambient air at sea level; 14.69 pounds per square inch. Air at sea level, 29.92 inches of mercury or 33.90 feet of water.

ATMOSPHERIC STILL
A refining vessel in which crude oil is heated and product is distilled off at the pressure of one atmosphere.

ATOMIZER, FUEL OIL
A nozzle or spraying device used to break up fuel oil into a fine spray so the oil may be brought into more intimate contact with the air in the combustion chamber. *See* Ultrasonic atomizer.

AUSTRALIAN OFFSET
A humorous reference to a well drilled miles away from proven production.

AUTOMATIC CUSTODY TRANSFER SYSTEM
A system of oil handling on a lease; receiving into tankage, measuring, testing, and turning into a pipeline the crude produced on a lease. Such automatic handling of oil is usually confined to leases with settled production.

AXLE GREASE
A cold-setting grease made of rosin oil, hydrated lime, and petroleum oils. *See* Grease.

B

BABBITT
A soft, silver-colored metal alloy of relatively low melting point used for engine and pump bearings; an alloy containing tin, copper, and antimony.

BACKFILL
To replace the earth dug from a ditch or trench; also, the earth removed from an excavation.

BACKHOE
A self-propelled ditching machine with a hydraulically operated arm equipped with toothed shovel that scoops earth as the shovel is pulled back toward the machine.

BACK-IN FARM-OUT
A farm-out agreement (q.v.) in which a retained non-operating interest of the lessor may be converted, at a later date, into a specified individual working interest (q.v.).

BACK OFF
To raise the drill bit off the bottom of the hole; to slack off on a cable or winch line; to unscrew.

BACK-PRESSURE VALVE
A Check Valve (q.v.)

BACKSIDE PUMPING
See Pumping, Backside

BACK-UP MAN
The person who holds (with a wrench) one length of pipe while another length is being screwed into or out of it.

BACKWASHING
Reversing the fluid flow through a filter to clean out sediment that has clogged the filter or reduced its efficiency. Backwashing is done on closed-system filters and on open-bed, gravity filters.

BAD OIL
Cut oil (q.v.)

BAFFLES
Plates or obstructions built into a tank or other vessel that change the direction of the flow of fluids or gases.

BAIL
To evacuate the liquid contents of a drill hole with the use of a long, cylindrical bucket (bailer).

BAIL DOWN
To reduce the level of liquid in a well bore by bailing.

BAILER
A cylindrical, bucket-like piece of equipment used in cable-tool drilling to remove mud and rock cuttings from the bore hole.

BAILER DART
The protruding "tongue" of the valve on the bottom of a bailer. When the dart reaches the bottom of the hole, it is thrust upward opening the valve to admit the mud-water slurry.

BAIT BOX
A pipeliner's lunch pail.

BALL AND SEAT
A type of valve used in a plunger pump.

BALLING OF THE BIT
The fouling of a rotary drilling bit in sticky, gumbo-like shale which causes a serious drag on the bit and loss of circulation.

BALL JOINT
A connector in a subsea, marine riser assembly whose ball and socket design permits an angular deflection of the riser pipe caused by horizontal movement of the drillship or floating platform of 10° or so in all directions.

BANDWHEEL
In a cable tool rig, the large vertical wheel that transmits power from the drilling engine to the crank and pitman assembly that actuates the walking beam. Used in former years in drilling with cable tools. Old pumping wells still use a bandwheel.

BAREFOOT COMPLETION
Wells completed in firm sandstone or limestone that show no indication of caving or disintegrating may be finished "barefoot," i.e., without casing through the producing interval.

BARITE
A mineral used as weighting material in drilling mud; a material to increase the density or weight per gallon or cubic foot of mud.

BARKER
A whistle-like device attached to the exhaust pipe of a one-cylindered oil

field engine so that the lease pumper can tell from a distance whether or not the engine is running. The noise the device makes resembles the bark of a hoarse fox.

BAROID
A specially processed barite (barium sulfate) to which Aquagel has been added, used as a conditioning material in drilling mud in order to obtain satisfactory cores and formation samples.

BARREL
(1) Petroleum barrel; a unit of measure for crude oil and oil products equal to 42 U.S. gallons. (2) Pump barrel; cylindrical body of an oil well pump.

BARREL HOUSE
A building on the refinery grounds where barrels are filled with various grades of lubricating and other oils, sealed, and made ready for shipment; oil house. See Drum.

BARREL MILE
The cost to move a barrel of oil or an equivalent amount of product one mile.

BASEMENT ROCK
Igneous or metamorphic rock lying below the sedimentary formations in the earth's crust. Basement rock does not contain petroleum deposits.

BASIC SEDIMENT
Impurities and foreign matter contained in oil produced from a well. See BS&W.

BASIN
A synclinal structure in the subsurface, once the bed of a prehistoric sea. Basins, composed of sedimentary rock, are regarded as good prospects for oil exploration.

BASTARD
(1) Any non-standard piece of equipment. (2) A kind of file. (3) A word used in grudging admiration, or as a term of opprobrium.

BATCH
A measured amount of oil or refined product in a pipeline or a tank; a shipment of oil or product by pipeline.

BATCHING SPHERE
An inflated, hard rubber sphere used in product pipelines to separate "incompatible" batches of product being pumped one behind the other. Fungible (q.v.) products are not physically separated, but gasoline is separated from diesel fuel and heating oils by batching spheres.

BATHOLITH
A great mass of intruded igneous or metamorphosed rock found at or near the surface of the earth. The presence of a batholith, often referred to as a shield, usually precludes drilling for oil or gas as there are no sedimentary formations above it. The largest batholith in the U.S. is in Idaho, underlying nearly two thirds of the state.

BATTERY
Two or more tanks connected together to receive oil production on a lease; tank battery.

BAUME, ANTOINE
The French chemist who devised a simple method to measure the relative weights of liquids using the hydrometer (q.v.).

BCD
Barrels per calendar day. *See* Stream day.

BEAD
A course of molten metal laid down by a welder (electric or oxyacetylene) in joining two pieces of metal. *See* Pipeline welding.

BEAKER SAMPLER
A metal or glass container with a small opening fitted with a stopper that is lowered into a tank of oil to obtain a sample of oil.

BEAM WELL
A well whose fluid is being lifted by rods and pump actuated by a walking beam (q.v.).

BEAN
A choke used to regulate the flow of fluid from a well.

BEAN JOINT
In early pipeline parlance, the joint of line pipe laid just before the break for lunch. When the bean joint was bucked in (q.v.) the pipeliners grabbed lunch buckets from the gang truck and found a comfortable place to eat.

BEARING, OUTBOARD
A shaft-supporting bearing outside the body or frame of a pump's gear box or engine's crankcase; a bearing on a pump's pinion shaft outside the gear box; a line-shaft bearing.

BEARING, SADDLE
A type of bearing for the support of a heavy, slow-moving member; e.g. the wide bearing on the samson post that supports the well's walking beam as it oscillates or rocks up and down.

BEARING, STIRRUP
A bearing and its frame in the shape of a saddle stirrup; e.g. the bearing connecting the pitman and the walking beam on an early-day cable tool drilling or pumping well.

BELL AND SPIGOT JOINT
A threaded pipe joint; the spigot or male end is threaded and screwed into the bell or female coupling. The female end of a coupling has threads on the inside circumference.

BELL HOLE
An excavation dug beneath a pipeline to provide room for the use of tools by workers; a hole larger in diameter at the bottom than at the top.

BELL-HOLE WELDER
A welder who can do oxyacetylene or electric welding in a bell hole. This requires a great deal of skill as the molten metal from the welding rod is being laid on upside down and tends to fall away from the weld; a skilled welder.

BELL NIPPLE
A large swage nipple for attaching casinghead fittings to the well's casing above the ground or at the surface. The bell nipple is threaded on the casing end and has a plain or weld-end to take the casinghead valves.

BELT HALL
A wooden shed built to protect the wide belt that runs from the engine to the bandwheel on a cable tool rig, or an old beam pumping well. The belt hall extends from the engine house to the derrick.

BENCH-SCALE TEST
Testing of methods or materials on so small a scale that it can be carried out on a laboratory table or specially constructed bench.

BENZINE
An old term for light, petroleum distillates in the gasoline and naphtha range.

BENZOL
The general term which refers to commercial benzene which may contain other aromatic hydrocarbons.

BEVELING MACHINE
An oxyacetylene pipe-cutting machine. A device that holds an acetylene cutting torch so that the ends of joints of pipe may be trimmed off at an angle to the pipe's long axis. Line pipe is beveled in preparation for welding joints together.

B.H.P.
Brake horsepower (q.v.)

BIG SPROCKET, ON THE
Said of a person who is moving in influential circles or has suddenly gone from a small job to one of considerably larger responsibility; a big operator, often used perjoratively.

BIOCHEMICAL CONVERSION
The use of bacteria to separate kerogen from oil shale. Certain bacteria will biodegrade the minerals in oil shale releasing the kerogen from the shale in liquid or semi-liquid form. (From studies made by Dr. Ten Fu Yen and Dr. Milo D. Appleman, University of Southern California, Professors of Bacteriology).

BIRD CAGE
(1) To flatten and spread the strands of a cable or wire rope. (2) The slatted or mesh-enclosed cage used to hoist workmen from crew boats to offshore platforms.

BIRD DOG
To pay close attention to a job or to follow a person closely with the intent to learn or to help; to follow up on a job until it is finished.

BIT
The cutting or pulverizing tool or head attached to the drill pipe in boring a hole in underground formations. *See* Drill bit.

BITUMINOUS SAND
Tar sand; a mixture of asphalt and loose sand which, when processed, may yield as much as 12 percent asphalt.

BLACK OILS
(1) A term denoting residual oil; oil used in ships' boilers or in large heating or generating plants; bunkers. (2) Black-colored oil used for lubricating heavy, slow-moving machinery where the use of higher-grade lubes would be impractical.

BLEED
To draw off a liquid or gas slowly. To reduce pressure by allowing fluid or gas to escape slowly; to vent the air from a pump.

BLEEDER VALVE
A small valve on a pipeline, pump, or tank from which samples are drawn or to vent air or oil; sample valve.

BLEEDING
The tendency of a liquid component to separate from a lubricant, as oil from a grease; to seep out.

BLEEDING CORE
A core sample of rock highly saturated and of such good permeability that oil drips from the core.

BLEED LINE
A line on the wellhead or blowout preventer stack through which gas pressure can be bled to prevent a threatened blowout.

BLENDING
The process of mixing two or more oils having different properties to obtain a lubricating oil of intermediate or desired properties. Certain classes of lube oils are blended to a specified viscosity. Other products, notably gasolines, are also blended to obtain desired properties.

BLIND FLANGE
A companion flange with a disc bolted to one end to seal off a section of pipe.

BLOCK
(1) A pulley or sheave in a rigid frame. (2) To prevent the flow of liquid or gas in a line. (3) A chock.

BLOCK AND TACKLE
An arrangement of ropes and blocks (pulleys) used to hoist or pull.

BLOCK GREASE
A grease of high melting point that can be handled in block or stick form. Block grease is used on large, slow-moving machinery, on axles and crude bearings. In contact with a hot journal bearing, the grease melts slowly lubricating the bearing.

BLOCKING
Pumping crude oil or refined products in batches or blocks through a pipeline.

BLOCK VALVE
A large, heavy-duty valve on a crude oil or products trunk line placed on each side of a pipeline river crossing to isolate possible leaks at the crossing.

BLOOM
The irridescent cast of color in lubricating oil.

BLOWBY
The escape of combustion or unburned fuel past the engine's piston and piston rings into the crankcase. Blowby occurs during the power stroke but unburned fuel can also escape during the compression stroke on spark-ignition engines.

BLOWDOWN STACK
A vent or stack into which the contents of a processing unit are emptied when an emergency arises. Steam is injected into the tank to prevent ignition of volatile material or a water quench is sometimes used.

BLOWING A WELL
Opening a well to let it blow for a short period to free the well tubing or casing of accumulations of water, sand, or other deposits.

BLOWING THE DRIP
To open the valve on a drip (q.v.) to drain off the "drip gasoline" and to allow the natural gas to "blow" for a moment to clear the line and drip off all liquid.

BLOWOUT
Out of control gas and/or oil pressure erupting from a well being drilled; a dangerous, uncontrolled eruption of gas and oil from a well; a wild well.

BLOWOUT PREVENTER
A stack or an assembly of heavy-duty valves attached to the top of the casing to control well pressure; a Christmas tree (q.v.).

BLOWPIPE (Welding and Cutting)
See Welding torch

BNOC
British National Oil Corporation (q.v.).

BOBTAIL
A short-bodied truck

BOB-TAIL PLANT (Gas)
A gas plant which extracts liquid hydrocarbons from the natural gas but does not break down the liquid product into its separate components.

BODY
Colloquial term for the viscosity of an oil.

BOILER HOUSE
(1) A lightly constructed building to house steam boilers; (2) To make a report without doing the work; to fake a report.

BOILOFF
The vaporization or gasification of liquefied natural gas (LNG) or other gases liquefied by applying high pressure and severe cooling. Boiloff occurs when the holding vessel's insulation fails to maintain the low temperature required to keep the gas in liquid form. Boiloff is a problem for shippers of LNG in the specially built ocean carriers.

BOLL WEEVIL
An inexperienced worker or "green hand" on a drilling crew.

BOLSTER
A support on a truck bed used for hauling pipe. The heavy wooden or metal beam rests on a pin that allows the forward end of the load to pivot as the truck turns a corner.

BONNET
The upper part of a gate valve that encloses the packing gland and supports the valve stem.

BONUS
Usually, the bonus is the money paid by the lessee for the execution of an oil and gas lease by the landowner. Another form is called an oil or royalty bonus. This may be in the form of an overriding royalty reserved to the landowner in addition to the usual one-eighth royalty.

BOOK VALUE
The worth of an oil company; properties and all facilities, less depreciation.

BOOM
A beam extending out from a fixed foundation or structure for lifting or hoisting; a movable arm with a pulley and cable at the outer end for hoisting or exerting tension on an object. *See* Boom-Cats.

BOOM-CATS
Caterpillar tractors equipped with side-booms and winches; used in pipeline construction to lift joints of pipe and to lower sections of the line into the ditch.

BOOMER
(1) A link and lever mechanism used to tighten a chain or cable holding a load of pipe or other material. (2) A worker who moves from one job to another.

BOOSTER STATION
A pipeline pumping station usually on a main line or trunk line; an intermediate station; a field station that pumps into a tank farm or main station.

BOOT
A tall section of 12 or 14-inch pipe used as a surge column at a lease tank battery, down stream of the oil/gas separator. The column, 15 to 25 feet high, provides a hydrostatic head (q.v.) and permits the escape of gas which prevents gas or vapor locks in the pipeline gathering system.

BOP STACK
Blowout preventer stack (q.v.)

BORE AND STROKE
See Pump specifications

BOREHOLE
The hole in the earth made by the drill; the uncased drill hole from the surface to the bottom of the well.

BOTTLENECKING
The deformation of the ends of the casing or tubing in the hanger resulting from excessive weight of the string of pipe and the squeezing action of the slips.

BORING MACHINE
A power-driven, large-diameter augur used to bore under roads, railroads, and canals for the purpose of installing casing or steel conduits to hold a pipeline.

BOTTOM-HOLE CHOKE
A device placed at the bottom of the tubing to restrict the flow of oil or regulate the gas-oil ratio.

BOTTOM-HOLE LETTER
An agreement by which an operator, planning to drill a well on his own land, secures the promise from another to contribute to the cost of the well. In contrast to a Dry-hole Letter, the former requires payment upon completion of the well whether it produces or not. A Bottom-hole Letter is often used by the operator as security for obtaining a loan to finance the drilling of the well.

BOTTOM-HOLE PRESSURE
The reservoir or formation pressure at the bottom of the hole. If measured under flowing conditions, readings are usually taken at different rates of flow in order to arrive at a maximum productivity rate. A decline in pressure indicates the amount of depletion from the reservoir.

BOTTOM-HOLE PUMP
A pump located in the bottom of the well and not operated by sucker rods and surface power unit. Bottom-hole pumps are compact, high volume units driven by an electric motor or hydraulically operated.

BOTTOM OUT
To reach total depth, to drill to a specified depth.

BOURDON TUBE
A small, crescent-shaped tube closed at one end, connected to a source of gas pressure at the other, used in pressure recording devices or in pilot-operated control mechanisms. With increases in gas pressure the Bourdon tube flexes (attempts to straighten) and this movement, through proper linkage, actuates recording instruments.

BOWL
A device that fits in the rotary table and holds the wedges or slips that support a string of tubing or casing.

BOX AND PIN JOINT
A type of screw coupling used to connect sucker rods and drill pipe. The box is a thick-walled collar with threads on the inside; the pin is threaded on the outer circumference and is screwed into the box.

BOYLE'S LAW
"The volume of any weight of gas is inversely proportional to the absolute pressure, provided the temperature remains constant."

BRADENHEAD GAS
Casinghead gas. Bradenhead was an early-day name for the wellhead or casinghead.

BRAKE HORSEPOWER (B.H.P.)
The power developed by an engine as measured at the drive-shaft; the actual or delivered horsepower as contrasted to "indicated horsepower" (q.v.).

BREA
A viscous, asphaltic material formed at oil seepages when the lighter fractions of the oil have evaporated, leaving the black, tar-like substance.

BREAK CIRCULATION
To resume the movement of drilling fluid down the drill pipe, through the "eyes" of the bit, and upward through the annulus to the surface.

BREAK TOUR
To begin operating 24 hours a day on three eight-hour shifts after rigging up on a new well. Until the derrick is in place and rigged up, mud pits dug, pipe racked, and other preparatory work done, the drill crew works a regular eight-hour day. When drilling commences, the crews break tour and begin working the three, eight-hour tours.

BREAKING DOWN THE PIPE
Unscrewing stands of drill pipe in one-joint lengths usually in preparation for stacking and moving to another well location.

BREAKOUT
(1) To isolate pertinent figures from a mass of data; to retrieve relevant information from a comprehensive report. (2) To loosen a threaded pipe joint.

BREAKOUT TANKAGE
Tankage at a take-off point or delivery point on a large crude oil or products pipeline.

BREATHING
The movement of oil vapors and air in and out of a storage tank owing to the alternate heating by day and cooling by night of the vapors above the oil in the tank.

BRIDGE OVER
The collapse of the walls of the bore hole around the drill column.

BRIDGE PLUG
An expandable plug used in a well's casing to isolate producing zones or to plug back to produce from a shallower formation; also to isolate a section of the bore hole to be filled with cement when a well is plugged.

BRIDLE
A sling made of steel cable fitted over the "horsehead" on a pumping jack and connected to the pump rod; the cable link between "horsehead" and pump rod on a pumping well.

BRIGHT STOCKS
High viscosity, fully refined, and de-waxed lubricating oils; used for blending with lower viscosity oils. The name originated from the clear, bright appearance of the de-waxed lubes.

BRING BOTTOMS UP
To wash rock cuttings from the bottom of the hole to the surface by maintaining circulation after halting the drilling operation. This allows time for the closer inspection of the cuttings and for a decision as to how to proceed when encountering a certain formation.

BRITISH NATIONAL OIL CORPORATION
The United Kingdom government agency that "participates" in drilling and production activities in the British sectors of the North Sea with U.S. oil companies and others; the "corporation" through which Britain assumes ownership of the U.K.'s share of the North Sea oil.

BROKE OUT
To be promoted; "He broke out as a driller at Midland;" to begin a new job after being promoted.

BRONC
A new driller promoted from helper; a new tool pusher up from driller; any newly promoted oil field worker whose performance is still untried.

BRUCKER SURVIVAL CAPSULE
A patented, self-contained survival vessel that can be lowered from an offshore drilling platform or semi-submersible in the event of a fire or other emergency. The vessel, of spheroid shape, is self-propelled and is equipped with first aid and life-support systems. Some models can accommodate 26 persons. *See* Whitaker system.

BSD
Barrels per stream-day. *See* Stream day.

B S & W
Short for basic sediment and water often found in crude oil.

BTX
Benzene - toluene - xylene; basic aromatics used in the manufacture of paints, synthetic rubber, agricultural chemicals, and chemical intermediates. The initials are used by refinery men in designating a unit of the refinery.

BUBBLE POINT
The pressure at which gas, held in solution in crude oil, breaks out of solution as free gas; saturation pressure.

BUBBLE POINT PUMP
A type of down hole oil pump very sensitive to gas. When the saturation pressure is reached, gas is released which gas-locks the pump until pressure is again built up by the oil flowing into the well bore. This type of pump regulates, in effect, oil production from a reservoir with a gas drive.

BUBBLETOWER
Any of the tall cylindrical towers at an oil refinery. *See* Fractionator.

BUCK UP
To tighten pipe joints with a wrench.

BUCKING THE TONGS
Working in a pipeline gang laying screw pipe; hitting the hooks (q.v.).

BULK PLANT
A distribution point for petroleum products. A bulk plant usually has tank car unloading facilities and warehousing for products sold in packages or in barrels.

BULL GANG
Common laborers who do the ditching and other heavy work on a pipeline construction job.

BULL PLUG
A short, tapered pipe fitting used to plug the open end of a pipe or throat of a valve.

BULL WAGON
A casing wagon (q.v.).

BULL WHEELS
On a cable-tool rig, the large wheels and axle located on one side of the derrick floor used to hold the drilling line. *See* Calf wheels.

BUMP-OFF A WELL
To disconnect a rod-line well from a central power unit.

BUMPER SUB
A slip-joint that is part of the string of drill pipe used in drilling from a drill ship to absorb the vertical motion of the ship caused by wave action. The slip joint is inserted above the heavy drill collars in order to maintain the weight of the collars on the drill bit as the drill pipe above the slip joint moves up and down with the motion of the ship.

BUMPER SUB (FISHING)
A hydraulically-actuated tool installed in the fishing string above the fishing tool to produce a jarring action. When the fishing tool has a firm hold on the lost drill pipe or tubing, which may also be stuck fast in the hole, the bumper sub imparts a jarring action to help free the "fish."

BUNKER "C" FUEL OIL
A heavy, residual fuel oil used in ships' boilers and large heating and generating plants.

BUNKERING
To supply fuel to vessels for use in the ships' boilers; the loading of bunker fuel on board ship for use by the ship's boilers.

BUNKHOUSE
Crew quarters, usually a portable building used on remote well locations to house the drilling crew and for supplies; quarters for single, oil field workers in the days when transportation to a nearby town was primitive or unavailable.

BURNER
A device for the efficient combustion of a mixture of fuel and air. *See* Ultrasonic atomizer.

BURN PIT
An excavation in which waste oil and other material is burned.

BURNING POINT
The lowest temperature at which a volatile oil in an open vessel will continue to burn when ignited by a flame. This temperature determines the degree of safety with which kerosine and other illuminates may be used.

BURTON, WILLIAM M.
The petroleum chemist who developed the first profitable means of cracking low-value middle distillates into lighter fractions (gasolines) by the use of heat and pressure.

BUTANE
A hydrocarbon fraction; at ordinary atmospheric conditions, butane is a gas but it is easily liquefied; one of the most useful LP-gases, widely used household fuel.

BUTTERFLY VALVE
A type of quick-opening valve whose orofice is opened and closed by a disk that pivots on a shaft in the throat of the valve.

BUTT-WELDED PIPE
Pipe made from a rectangular sheet of steel which is formed on mandrels. The two edges of the sheet are butted together and welded automatically.

BUY-BACK CRUDE OIL
In foreign countries, buy-back oil is the host government's share of "participation crude" it permits the company holding the concession (the producer) to buy back. This occurs when the host government has no market for its share of oil received under the joint-interest or participation agreement.

BYPASS VALVE
A valve by which the flow of liquid or gas in a system may be shunted past a part of the system through which it normally flows; a valve that controls an alternate route for liquid or gas.

C

CABLE TOOLS
The equipment necessary for cable-tool drilling of a well. A heavy metal bar, sharpened to a chisel-like point on the lower end is attached to a drilling rope or wire line (cable) which is fastened to a walking beam above the rig floor that provides an up and down motion to the line and the metal drilling tool. The drilling tool, called a bit, comes in a variety of cutting-edge configurations.

CALF WHEELS
The spool or winch located across the derrick floor from the bull wheels (q.v.) on a cable tool rig. The casing is usually run with the use of the calf wheels which are powered by the bandwheel (q.v.). A line from the calf wheels runs to the crown block and down to the rig floor.

CALIPER LOG
A tool for checking casing downhole for any bending or flattening or other deformation prior to running and setting a packer or other casing hardware. *See* Drift mandrel.

CAMP, COMPANY
A small community of oil field workers; a settlement of oil company employees living on a lease in company housing. In the early days, oil companies furnished housing, lights, gas, and water free or at a nominal charge to employees working on the lease and at nearby company installations—pumping stations, gasoline plants, tank farms, loading racks, etc. Camps were known by company lease or simply the lease name, e.g. Gulf Wolf Camp, Carter Camp, and Tom Butler.

CANNING LINE
A facility at refinery where cans are filled with lubricating oil, sealed, and put in cases. Modern canning lines are fully automated.

CAPITAL ASSETS
Assets acquired for investment and not for sale, and requiring no personal services or management duties. In Federal income tax law, oil and gas leases are, ordinarily, property used in the taxpayers trade or business and are not capital assets. Royalty, if held for investment, is usually considered a capital asset.

CAPITAL EXPENDITURES
Non-deductible expenditures which must be recovered through depletion or depreciation. In the oil industry, these items illustrate expenditures that must be capitalized: geophysical and geologic costs, well equipment, and lease bonuses paid by lessee.

CAPITAL-GAP DILEMMA
The growing disproportion of capital investment to oil reserves discovered; the increasing need for investment capital coupled with diminishing results in terms of oil and gas discovered; spending more to find less oil.

CAPITAL INTENSIVE INDUSTRY
Said of the oil industry because of the great amounts of investment capital required to search for and establish petroleum reserves.

CAPITAL STRING
Another name for the production string (q.v.).

CAPPING
Closing in a well to prevent the escape of gas or oil.

CAPTURED BOLT
A bolt held in place by a fixed nut or threaded piece. The bolt can be

tightened or loosened but can not be removed completely because of a shoulder at the end of the bolt. Captured bolts are in reality a part of an adjustable piece and are so made to preclude the chance of being removed and dropped or because of limited space and accessibility in an item of equipment.

CARBON BLACK
A fine, bulky carbon obtained as soot by burning natural gas in large horizontal "ovens" with insufficient air.

CARBON PLANT
A plant for the production of carbon black by burning natural gas in the absence of sufficient air. Carbon plants are located close to a source of gas and in more or less isolated sections of the country because of the heavy emission of smoke.

CASING
Steel pipe used in oil wells to seal off fluids from the bore hole and to prevent the walls of the hole from sloughing off or caving. There may be several strings of casing in a well, one inside the other. The first casing put in a well is called surface pipe which is cemented into place and serves to shut out shallow water formations and also as a foundation or anchor for all subsequent drilling activity. See Production string.

CASING PRESSURE
Pressure between the casing and the well's tubing.

CASING, SHALLOW-WELL
Small-diameter casing of lighter weight than conventional casing used in deep wells. The lighter-weight casing is less costly, easier to handle, and adequate for certain kinds of shallow, low-pressure wells.

CASING SHOE
A reinforcing collar of steel screwed onto the bottom joint of casing to prevent abrasion or distortion of the casing as it forces its way past obstruction on the wall of the bore hole. Casing shoes are about an inch thick and 10 to 16 inches long and are an inch or so larger in diameter in order to clear a path for the casing.

CASINGHEAD
The top of the casing set in a well; the part of the casing that protrudes above the surface and to which the control valves and flow pipes are attached.

CASINGHEAD GAS
Gas produced with oil from an oil well as distinguished from gas from a gas well. The casinghead gas is taken off at the top of the well or at the separator.

CASINGHEAD GASOLINE
Liquid hydrocarbons separated from casinghead gas by the reduction of pressure at the well head or by a separator, or absorption plant. Casinghead gasoline or natural gasoline, is a highly volatile, water-white liquid.

CASING WAGON
A small, low cart for moving casing from the pipe rack to the derrick floor. Two wagons are used. The forward wagon holds the pipe in a V-shaped cradle; the rear wagon is in reality a lever on wheels which raises the end of the casing so it is free to be pulled.

CAT
Short for Caterpillar tractor, a crawler-type tractor that moves on metal tracks made in segments and connected with pins to form an "endless" tread.

CATALYSIS
A process in which the rate of chemical reaction is hastened or retarded by contact with an unrelated substance called a catalyst.

CATALYST
A substance which hastens or retards a chemical reaction without undergoing a chemical change itself during the process.

CAT CRACKER
A large refinery vessel for processing reduced crude oil, naphthas, or other intermediates in the presence of a catalyst. *See* Fluid catalytic cracking unit.

CATENARY
The sag or curve of a cable or chain stretched between two supports.

CATHEAD
A spool-shaped hub on a winch shaft around which a rope is wound for pulling and hoisting; a power take-off spool used by the driller as he operates the cat line (q.v.).

CATHODIC PROTECTION
An anti-corrosion technique for metal installations—pipelines, tanks, buildings—in which weak electric currents are set up to offset the current associated with metal corrosion. Carbon or non-ferrous anodes (q.v.) buried near the pipeline are connected to the pipe. Current flowing from the corroding anode to the metal installation control the corrosion of the installation.

CAT LINE
A hoisting or pulling rope operated from a cathead (q.v.). On a drilling rig, the rope used by the driller to exert a pull on pipe tongs in tightening (making up) or loosening (breaking out) joints of pipe.

CATTLE GUARD
A ground-level, trestle-like crossing placed at an opening in a pasture fence to prevent cattle from getting out while permitting vehicles to cross over the metal or wooden open framework.

CAT WALK
A raised, narrow walkway between tanks or other installations.

CAVEY FORMATION
A formation that tends to cave or slough into the well's bore hole. In the parlance of cable-tool drillers, "the hole doesn't stand up."

CAVITATION
The creation of a partial vacuum or a cavity by a high-speed impeller blade or boat propeller moving in or through a liquid. Cavitation is also caused by a suction pump drawing in liquid where there is an insufficient suction or hydrostatic head to keep the line supplied.

CD
Contract depth; the depth of a well called for or specified in the drilling contract.

CELLAR
An excavation dug at the drill site before erecting the derrick to provide working space for the casinghead equipment beneath the derrick floor. Blowout preventer valves (BOP stack) are also located beneath the derrick floor in the cellar.

CELLAR DECK
Lower deck on a large, double-decked, semi-submersible drilling platform.

CEMENT, TO
To fix the casing firmly in the hole with cement, which is pumped through the drill pipe to the bottom of the casing and up into the annular space between the casing and the walls of the well bore. After the cement sets (hardens) it is drilled out of the casing. The casing is then perforated to allow oil and gas to enter the well.

CEMENT SLURRY
A soupy mixture of water (or other liquid) and cement. Slurries are thin so they can be pumped and to enable the cement to penetrate cracks and crevices and to fill all voids.

CEMENT SQUEEZE
A method whereby perforations, large cracks, and fissures in the wall of the bore hole are forced full of cement and sealed off.

CENTRAL POWER
A well-pumping installation consisting of an engine powering a large-diameter, horizontal band wheel with shackle-rod lines attached to its circumference. The band wheel is an excentric and as it revolves on a vertical axle a reciprocating motion is imparted to the shackle rods. A central power may pump from 10 to 25 wells on a lease.

CENTRIFUGAL PUMP
A pump made with blades or impellers in a close-fitting case. The liquid is pushed forward by the impellers as they rotate at high speed. Centrifugal pumps, because of their high speed, are able to handle large volumes of liquid.

CENTRIFUGE
A motor-driven machine in which samples of oil or other liquids are rotated at high speed causing suspended material to be forced to the bottom of a graduated sample tube so that the percent of impurities or foreign matter may be observed. Some centrifuges are hand-operated. *See* Shake out.

CFM
Cubic feet per minute

CHAIN TONGS
A pipe wrench with a flexible chain to hold the toothed wrench-head in contact with the pipe. The jointed chain can be looped around pipes of different diameters and made fast in dogs on the wrench head.

CHANNEL
A "vacation" or void in a cement squeeze job allowing salt water or other fluid into the production zone or another interval in the annular space. Also, in waterflooding, a natural void or "path" in a formation permitting the injection fluid to break through to a producing well from the injection well subverting the waterflooding project. *See* Squeezing a well.

CHARCOAL TEST
A test to determine the gasoline content of natural gas.

CHARGING STOCK
Oil that is to be "charged" or treated in a particular refinery unit.

CHASE THREADS
To straighten and clean threads of any type.

CHEATER
A length of pipe used to increase the leverage of a wrench; anything used to lengthen a handle to increase the applied leverage.

CHECK VALVE
A valve with a free-swinging tongue or clapper that permits fluid in a pipeline to flow in one direction only; back-pressure valve.

CHEMICAL FEEDER PUMP
A small-volume pump used on oil leases to inject chemicals into flow lines. The pump may be located at the well head and be actuated by the motion of the pumping jack. The chemical is used to break down water/oil emulsions that may be contained in the crude oil stream.

CHILLERS
Refinery apparatus in which the temperature of paraffin distillates is lowered preparatory to filtering out the solid wax.

CHOCK
A wedge or block to prevent a vehicle or other movable object from shifting position; a chunk.

CHOKE
A heavy steel nipple inserted into the production tubing that closes off the flow of oil except through an orifice in the nipple. Chokes are of various sizes. It is customary to refer to the production of a well as so many barrels through (or on) a (e.g.) 22/64th-inch choke.

CHRISTMAS TREE
An assembly of valves mounted on the casinghead through which a well is produced. The Christmas tree also contains valves for testing the well and for shutting it in if necessary.

CHRISTMAS TREE (MARINE)
A subsea production system similar to a conventional land tree except it is assembled complete for remote installation on the sea floor with or without diver assistance. The marine tree is installed from the drilling platform; it is lowered into position on guide cables anchored to foundation legs implanted in the ocean floor. The tree is then latched mechanically or hydraulically to the casing head by remote control.

CHURN DRILLING
Another name for cable tool drilling because of the up and down, churning motion of the drill bit.

CID
Cubic inch displacement; the volume "swept out" or evacuated by the pistons of an engine in one working stroke; used to describe the size (and by implication, the power) of an automobile engine.

C.I. PLUG
A cast-iron plug; a flat plug used to close the end of a pipe or a valve.

CIRCLE JACK (CABLE-TOOL RIG)
A device used on the floor of a cable-tool rig to make up and break out (tighten and loosen) joints of drilling tools, casing or tubing; a jacking device operated on a toothed or notched metal, circular track placed around the pipe joint protruding from the bore hole, above the floor. The jack is operated manually with a handle, and is connected to a wrench which tightens the pipe joint as the jack is advanced, notch by notch.

CIRCULATE
To pump drilling fluid into the bore hole through the drill pipe and back up the annulus between the pipe and the wall of the hole; to cease drilling but to maintain circulation for any reason. When closer inspection of the formation rock just encountered is desired, drilling is halted as circulation is continued to "bring bottoms up" (q.v.).

CIRCULATION
The round trip made by drilling mud; down through the drill pipe and up on the outside of the drill pipe, between the pipe and the walls of the bore hole. If circulation is "lost," the flow out of the well is less than the flow into the well; the mud may be escaping into some porous formation or a cavity down hole. *See* Lose returns.

CLAPPER
The internal moving part, the "tongue" of a check valve that permits a liquid or gas to flow in one direction only in a pipeline. Like a trap door, the check-valve clapper works on a hinge attached to the body of the valve. When at rest the clapper is a few degrees off the vertical or, as in certain valves, completely horizontal.

CLAY
The filtering medium, especially Fuller's earth, used in refining; a substance which tends to adsorb the coloring materials present in oil which passes through it.

CLAY PERCOLATER
Refinery filtering equipment employing a type of clay to remove impurities or to change the color of lubricating oils.

CLEAN CARGO
Refined products—distillates, kerosine, gasoline, jet fuel—carried by tankers, barges, and tank cars; all refined products except bunker fuels and residuals (q.v.).

CLEAN OIL
Crude oil containing less than one percent sediment and water (BS&W); pipeline oil; oil clean enough to be accepted by a pipeline for transmission.

CLEAN-OUT BOX
A square or rectangular opening on the side of a tank or other vessel through which the sediment that has accumulated can be removed. The opening is closed with a sheet of metal (a door) bolted in place.

CLEAN-UP TRIP
Running the drill pipe into the hole for circulation of mud only; to clean the bore hole of cuttings.

CLEVIS
A U-shaped metal link or shackle with the ends of the U drilled to hold a pin or bolt; used as a connecting link for a chain or cable.

CLOSE NIPPLE
A very short piece of pipe having threads over its entire length; an all-thread nipple.

CLOSED-IN
Refers to a well, capable of producing, that is shut in (q.v.).

CLOUD POINT
The temperature at which paraffin wax begins to crystallize or separate from the solution, imparting a cloudy appearance to the oil as it is chilled under prescribed conditions.

CO_2 INJECTION
A secondary recovery technique in which carbon dioxide is injected into service wells in a field as part of a miscible recovery program. CO_2 is used in conjunction with water flooding.

CO_2-SHIELDED WELDING
See Welding, CO_2 shielded

COAL GASIFICATION
A process for producing "natural gas" from coal. Coal is heated and brought in contact with steam. Hydrogen atoms in the vapor combine with coal's carbon atoms to produce a hydrocarbon product similar to natural gas.

COAL OIL
Kerosine made from distilling crude oil in early-day pot stills; illuminating and heating oil obtained from the destructive distillation of bituminous coal.

COATING & WRAPPING
A field operation in preparing a pipeline to be put in the ditch (lowered in). The line is coated with a tar-like substance and then spiral-wrapped with tough, chemically impregnated paper. Machines that ride the pipe coat

and wrap in one continuous operation. Coating and wrapping protects the pipeline from corrosion. For large pipeline construction jobs, the pipe may be coated and wrapped at the mill or in yards set up at central points along the right-of-way.

COIL CAR (OR TRUCK)
A tank car or transport truck equipped with heating coils in order to handle viscous liquids that will not flow at ordinary temperatures.

COKE, PETROLEUM
Solid or fixed-carbon that remains in refining processes after distillation of all volatile hydrocarbons; the hard, black substance remaining after oils and tars have been driven off by distillation.

COKING
(1) The process of distilling a petroleum product to dry residue. With the production of lighter distillable hydrocarbons, an appreciable deposit of carbon or coke is formed and settles to the bottom of the still. (2) The undesirable building up of carbon deposits on refinery vessels.

COLD PINCH, TO
To flatten the end of a pipe with a hydraulically powered set of pinchers. Pinching the pipe-end is done to make a quick, temporary closure in the event a loaded pipeline is accidentally ruptured.

COLLAR
A coupling for two lengths of pipe; a pipe fitting with threads on the inside for joining two pieces of threaded pipe of the same size.

COLLAR CLAMP
A device fitted with rubber gaskets bolted around a leaking pipe collar. The clamp is effective in stopping small leaks but is used only as a temporary measure until permanent repairs can be made.

COLLAR POUNDER OR PECKER
A pipeline worker who beats time with a hammer on the coupling into which a joint of pipe is being screwed by a tong gang. The purpose is two fold: to keep the tong men pulling in unison and to warm up the collar so that a tighter screw-joint can be made.

COLLET CONNECTOR
A component of a subsea drilling system; a mechanically or hydraulically operated latching collar connecting the marine riser (q.v.) to the blowout preventer stack.

COME-ALONG
A lever and short lengths of chain with hooks attached to the ends of the chains; used for tightening or pulling a chain. The hooks are alternately moved forward on the chain being tightened.

COMMERCIAL WELL
A well of sufficient net production that it could be expected to pay out in a reasonable time and yield a profit for the operator. A shallow, 50-barrel-a-day well in a readily accessible location on shore could be a commercial well whereas such a well in the North Sea or in the Arctic Islands would not be considered commercial.

COMMINGLING
The intentional mixing of petroleum products having similar specifications. In some instances, products of like specification are commingled in a product pipeline for efficient and convenient handling; producing two pay zones in the same well bore.

COMMON CARRIER
A person or company having state or Federal authority to perform public transportation for hire; an organization engaged in the movement of petroleum products—oil, gas, refined products—as a public utility and common carrier.

COMPANION FLANGE
A two-part connector or coupling; one part convex, the other concave. The two halves are held together by bolts and nuts. This type flange or "union" is used on small-diameter piping.

COMPLETE A WELL
To finish a well so that it is ready to produce oil or gas. After reaching total depth (TD) casing is run and cemented; casing is perforated opposite the producing zone, tubing is run and control and flow valves are installed at the well head. Well completions vary according to the kind of well, depth, and the formation from which it is to produce.

COMPRESSER PLANT
A pipeline installation to pump natural gas under pressure from one location to another through a pipeline.

COMPRESSION CUP
A grease cup; a container for grease made either with a screw cap or spring-loaded cap for forcing the grease onto a shaft bearing.

COMPRESSION-IGNITION ENGINE
A diesel engine (q.v.)

COMPRESSION RATIO
The ratio of the volume of an engine's cylinder at the beginning of the compression stroke to volume at the end of the stroke. High compression engines are generally more efficient in fuel utilization than those with lower compression ratios.

CONCESSION
An agreement (usually with a foreign government) to permit an oil company to prospect for and produce oil in the area covered by the agreement.

CONDEMNATION
The taking of land by purchase, at fair market value, for public use and benefit by state or Federal government; as well as by certain other agencies and utility companies having power of eminent domain (q.v.).

CONDENSATE
Liquid hydrocarbons produced with natural gas which are separated from the gas by cooling and various other means. Condensate generally has an A.P.I. gravity of 50° to 120° and is water-white, straw or bluish in color.

CONDENSATE WATER
Water vapor in solution with natural gas in the formation. When the gas is produced the water vapor condenses into liquid as both pressure and temperature are reduced. *See* Retrograde Gas Condensate

CONDENSATION
The transformation of a vapor or gas to a liquid by cooling or an increase in pressure or both simultaneously.

CONDENSER
A water-cooled heat exchanger used for cooling and liquefying vapors.

CONDUCTOR CASING
A well's surface pipe used to seal off near-surface water, prevent the caving or sloughing of the walls of the hole, and as a conductor of the drilling mud through loose, unconsolidated shallow layers of sand, clays, and shales. *See* Casing.

CONE ROOF
A type of tank roof built in the form of a flat, inverted cone; an old-style roof for large crude storage tanks, but still employed on tanks storing less volatile products. *See* Floating roof.

CONFIRMATION WELL
A well drilled to "prove" the formation or producing zone encountered by an exploratory or wildcat well. *See* Step-out well.

CONGLOMERATE
A type of sedimentary rock compounded of pebbles and rock fragments of various sizes held together by a cementing material, the same type material that holds sandstone together. Conglomerates are a common form of reservoir rock.

CONICAL-TOWER PLATFORM

A type of offshore drilling platform made of reinforced concrete for use in arctic waters where pack ice prevents the use of conventional platform construction. The structure is a truncated cone supporting a platform from which the wells are drilled.

CONNATE WATER

The water present in a petroleum reservoir in the same zone occupied by oil and gas. Connate water is not to be confused with bottom or edge water. Connate water occurs as a film of water around each grain of sand in granular reservoir rock and is held in place by capillary attraction.

CONNECTION FOREMAN

The supervisor, the boss of a pipeline connection gang (q.v.).

CONNECTION GANG

A pipeline crew that lays field gathering lines, connects stock tanks to gathering lines, and repairs pipelines and field pumping units in their district. Connection gangs also install manifolds and do pipe work in and around pumping stations. A typical gang of 8 or 10 men has a welder and a helper, a gang-truck driver and swamper (helper), 3 or 4 pipeliners, and a connection foreman.

CONSORTIUM

An international business association organized to pursue a common objective; (e.g.) to explore, drill, and produce oil.

CONSUMER GAS

Gas sold by an interstate gas pipeline company to a utility company for resale to consumers.

CONTOUR LINE

A line (as on a map) connecting points on a land surface that have the same elevation above or below sea level.

CONTOUR MAP

A map showing land surface elevations by the use of contour lines (q.v.). Structure contour maps are used by geologists and geophysicists to depict subsurface conditions or formations. See Isopachous map.

CONTROL PANEL

An assembly of indicators and recording instruments—pressure gauges, warning lamps, and other digital or audio signals—for monitoring and controlling a system or process.

COOLING TOWER

A large louvered structure (usually made of wood) over which water flows to cool and aerate it. Although most cooling towers are square or rectangular in shape, some are cylindrical, open at the bottom and top, which produces strong air currents through the center of the structure for more rapid cooling.

CORD ROAD
A passable road made through a swampy, boggy area by laying logs or heavy timbers side by side to make a bumpy but firm surface; a log road.

CORE BARREL
A device with which core samples of rock formations are cut and brought to the surface; a tube with cutting edges on the bottom circumference, lowered into the well bore on the drill pipe and rotated to cut the plug-like sample.

CORE BIT
A special drill bit for cutting and removing a plug-shaped rock sample from the bottom of the well bore.

CORE RECORD
A record showing the depth, character, and fluid content of cores taken from a well.

CORE SAMPLE
A solid column of rock, usually from two to four inches in diameter, taken from the bottom of a well bore as a sample of an underground formation. Cores are also taken in geological studies of an area to determine its oil and gas prospects.

CORRELATIVE RIGHTS, DOCTRINE OF
The inherent right of an owner of oil or gas in a field to his share of the "reservoir energy", and his right to be protected from wasteful practices by others in the field.

COST CRUDE OIL
Crude oil produced from an operator's own wells; oil produced at "cost" on a lease or concession acreage as compared to purchased crude.

COUPLING
A collar; a short pipe fitting with both ends threaded on the inside circumference; used for joining two lengths of line pipe or casing or tubing.

COUPLING POLE
The connecting member between the front and rear axles of a wagon or four-wheel trailer. To lengthen the frame of the vehicle, a pin in the pole can be removed and the rear-axle yoke (which is fastened to the pole by the pin) moved back to another hole. On pipe-carrying oil field trailers, the coupling pole is a telescoping length of steel tubing. The trailer can be made as long as necessary for the load.

CRACK A VALVE
To open a valve so slightly as to permit a small amount of fluid or gas to escape.

CRACKING
The refining process of breaking down the larger, heavier, and more complex hydrocarbon molecules into simpler and lighter molecules. Cracking is accomplished by the application of heat and pressure, and in certain advanced techniques, by the use of a catalytic agent. Cracking is an effective process for increasing the yield of gasoline from crude oil.

CRACKING A VALVE
Opening a valve very slightly.

CRANK
An arm attached at right angles to the end of a shaft or axle for transmitting power to or from a connecting rod or pitman (q.v.).

CRATER
(1) A bowl-shaped depression around a blow-out well caused by the caving in and collapse of the surrounding earth structure. (2) To fail or fall apart (colloquial).

CROSSHEAD
A sliding support for a pump or compressor's connecting rod. The rod is attached to a heavy "head" which moves to and fro on a lubricated slide in the pump's frame. Screwed into the other end of the crosshead is the pump's piston rod or plunger rod. A crosshead moves back and forth in a horizontal plane or up and down in a vertical plane transmitting the power from the connecting rod to the pump's piston rod.

CROSSOVER
A stile; a step-and-platform unit to provide access to a work platform or an elevated crossing. See Stile.

CROWBAR CONNECTION
A humorous reference to an assembly of pipe fittings so far out of alignment that a crowbar is required to force them to fit.

CROWN BLOCK
A stationary pulley system located at the top of the derrick used for raising and lowering the string of drilling tools; the sheaves and supporting members to which the lines of the traveling block (q.v.) and hook are attached.

CROWN PLATFORM
A platform at the very top of the derrick that permits access to the sheaves of the crown block and provides a safe area for work involving the gin pole (q.v.).

CRUDE OIL
Oil as it comes from the well; unrefined petroleum.

CRUDE STILL
A primary refinery unit; a large vessel in which crude oil is heated and various components taken off by distillation.

CRUMB BOSS
A person responsible for cleaning and keeping an oil field bunkhouse supplied with towels,. bed linen, and soap; a construction camp house-keeper.

CRUMB OUT
To shovel out the loose earth in the bottom of a ditch; also, to square up the floor and side of the ditch in preparation for laying of pipe.

CRYOGENICS
A branch of physics that relates to the production and effects of very low temperatures. The process of reducing natural gas to a liquid by pressure and cooling to very low temperatures employs the principles of cryogenics.

CUBES
Short for cubic inch displacement; CID (q.v.).

CULTIVATOR WRENCH
Any square-jawed, adjustable wrench that is of poor quality or worn out. *See* Knuckle-buster.

CUP GREASE
Originally, a grease used in compression cups (q.v.) but today the term refers to grease having a calcium fatty-acid soap base. *See* Grease.

CUPS
Discs with edges turned at right angles to the body used on plungers in certain kinds of pumps; discs of durable plastic or other tough, pliable material used on pipeline pigs or scrapers to sweep the line.

CUT
(1) A hydrocarbon fraction, the result of a distillation process. (2) To dilute with a foreign substance.

CUT-OIL
Crude oil partially emulsified with water; oil and water mixed in such a way as to produce an emulsion in which minute droplets of water are encased in a film of oil. In such case the water, although heavier, can not separate and settle to the bottom of a tank until the mixture is heated or treated with a chemical. *See* Roll A Tank.

CUTTING OILS
Special oils used to lubricate and cool metal-cutting tools.

CUTTINGS
Chips and small fragments of rock as the result of drilling that are brought to the surface by the flow of the drilling mud as it is circulated. Cuttings are important to the geologist who examines them for information concerning the type of rock being drilled. See Sample.

CUTTING TORCH
A piece of oxyacetylene welding and cutting equipment; a hand-held burner to which the oxygen and acetylene hoses are attached. The gases when ignited by the welder's lighter produce a small, intense flame that "cuts" metal by melting it. See Welding torch.

CYCLING (OF GAS)
Return to a gas-reservoir of gas remaining after extraction of liquid hydrocarbons for the purpose of maintaining pressure in the reservoir, and thus increasing the ultimate recovery of liquids from the reservoir.

CYCLING PLANT
An oil field installation that processes natural gas from a field, strips out the gas liquids, and returns the dry gas to the producing formation to maintain reservoir pressure.

CYLINDER OIL
Oils used to lubricate the cylinders and valves of steam engines.

CYLINDER STOCK
A class of highly viscous oils so called because originally their main use was in preparation of products to be used for steam cylinder lubrication.

D

D&P PLATFORM
A drilling and production offshore platform.

DAMPING SUB
Essentially, a downhole "shock absorber" for a string of drilling tools; a 6 to 8-foot-long device, a part of the drill assembly, that acts to dampen bit vibration and impact loads during drilling operations. Damping subs are

of the same diameter as the drill pipe into which they are screwed to form a part of the drill string.

DEADMAN
(1) A substantial timber or plug of concrete buried in the earth to which a guy wire or line is attached for bracing a mast or tower; a buried anchoring piece capable of withstanding a heavy pull. (2) A land-side mooring device used with lines and cables when docking a vessel.

DEADMAN CONTROL
A device for shutting down an operation should the attendant become incapacitated. The attendant using such a device must consciously exert pressure on a hold-down handle or lever to work the job. When pressure is relaxed owing to some emergency, the operation will automatically come to a halt.

DEAD OIL
Crude oil containing no dissolved gas when it is produced.

DEAD WELL
A well that will not flow, and in order to produce must be put on the pump.

DEADWOOD
Material inside a tank or other vessel such as pipes, supports, and construction members that reduce the true volume of the tank by displacing some of the liquid contents.

DEFICIENCY GAS
The difference between a quantity of gas a purchaser is obligated by contract either to take or pay for if not taken, and the amount actually taken.

DEMULSIFIER
A chemical used to "break down" crude oil/water emulsions. The chemical reduces the surface tension of the film of oil surrounding the droplets of water. Thus freed, the water settles to the bottom of the tank.

DEMURRAGE
The charge incurred by the shipper for detaining a vessel, freight car, or truck. High loading rates for oil tankers are of utmost importance in order to speed turnaround and minimize demurrage charges.

DENSMORE, AMOS
The man who first devised a method of shipping crude oil by rail. In 1865 he mounted two "iron banded" wooden tanks on a railway flatcar. The tanks or tubs held a total of 90 barrels. Densmore's innovation was the forerunner of the "unit train" for hauling oil and products, and the latest development, Tank Train (q.v.).

DEPLETION ALLOWANCE
See Percentage depletion

DEPOSIT
An accumulation of oil or gas capable of being produced commercially.

DEPROPANIZER
A unit of a processing plant where propane, a liquid hydrocarbon, is separated from natural gas.

DERRICK
A wooden or steel structure built over a well site to provide support for drilling equipment and a tall mast for raising and lowering drill pipe and casing; a drilling rig.

DERRICK FLOOR
The platform (usually 10 feet or more above the ground) of a derrick on which drilling operations are carried on; rig floor.

DERRICKMAN
A member of the drilling crew who works up in derrick on the tubing board racking tubing or drill pipe as it is pulled from the well and unscrewed by other crew members on the derrick floor.

DERRICK, PUMPING
In the early days, before the widespread use of portable units for pulling and reconditioning a well, the original derrick, used for drilling, was often replaced by a smaller, shorter derrick called a pumping derrick or pumping rig. Well workovers could be done with these rigs, the well also could be pumped by pumping jack or by a walking beam.

DESALTING PLANT
An installation that removes salt water and crystalline salt from crude oil streams. Some plants use electrostatic precipitation, others employ chemical processes to remove the salt.

DETERGENT OILS
Lubricating oils containing additives that retard the formation of gums, varnishes, and other harmful engine deposits. The detergents act to keep all products of oxidation and other foreign matter in suspension which permits it to be removed by the engine's filtering system.

DETROIT IRON
A humerous reference to a large, old car or truck.

DEVELOPMENT
The drilling and bringing into production of wells in addition to the discovery well on a lease. The drilling of development wells may be required by the express or implied covenants of a lease.

DEVELOPMENT CLAUSE
The drilling and delay-rental clause of a lease; also, express clauses specifying the number of development wells to be drilled.

DEVELOPMENT WELLS
Wells drilled in an area already proved to be productive.

DEVIATED HOLE
A well bore that is off the vertical either by design or accident.

DEW POINT
The temperature at which water vapor condenses out of a gas at 14.7 psia, (pounds per square inch absolute) or at sea level.

DIESEL, RUDOLPH
The German mechanical engineer who invented the internal combustion engine that bears his name.

DIESEL ENGINE
A four-stroke cycle internal combustion engine that operates by igniting a mixture of fuel and air by the heat of compression, and without the use of an electrical ignition system.

DIESELING
The tendency of some gasoline engines to continue running after the ignition has been shut off. This is often caused by improper fuel or carbon deposits in the combustion chamber hot enough to ignite the gasoline sucked into the engine as it makes a few revolutions after being turned off.

DIFFERENTIAL-PRESSURE STICKING
Another name for wall sticking (q.v.), a condition downhole when a section of drill pipe becomes stuck or hung up in the deposit of filter cake on the wall of the bore hole.

DIGGER
One who digs or drills a well; a driller.

DIGGING TOOLS
Hand tools used in digging a ditch, i.e., shovels, picks, mattocks, spades.

DIP
The angle that a geological stratum makes with a horizontal plane (the horizon); the inclination downward or upward of a stratum or bed.

DIRECTIONAL DRILLING
The technique of drilling at an angle from the vertical by deflecting the drill bit. Directional wells are drilled for a number of reasons: to develop an offshore lease from one drilling platform; to reach a pay zone beneath land where drilling cannot be done; (e.g.) beneath a railroad, cemetery, a lake;

and to reach the production zone of a burning well to flood the formation. *See* Killer well.

DIRTY CARGO
Bunker fuel and other black residual oils.

DISCOVERY WELL
An exploratory well that encounters a new and previously untapped petroleum deposit; a successful wildcat well. A discovery well may also open a new horizon in an established field.

DISCOVERY WELL ALLOWABLE
An allowable above that of wells in a settled field. Some states allow the operators of a discovery well to produce at the maximum efficiency rate (MER) until the costs of the well have been recovered in oil or gas.

DISPATCHER
One who directs the movement of crude oil or product in a pipeline system. He receives reports of pumping rates, line pressures, and monitors the movement of batches of oil; he may also operate remote, unmanned stations.

DISPOSAL WELL
A well used for the disposal of salt water. The water is pumped into a subsurface formation sealed off from other formations by impervious strata of rock; a service well.

DISSOLVED GAS
Gas contained in solution with the crude oil in the reservoir. *See* Solution gas.

DISSOLVED-GAS DRIVE
The force of expanding gas dissolved in the crude oil in the formation that drives the oil to the well bore and up to the surface through the production string.

DISTILLATE
Liquid hydrocarbons, usually water-white or pale straw color, and of high API gravity (above 60°) recovered from wet gas by a separator that condenses the liquid out of the gas stream. *See* Condensate. (Distillate is an older term for the liquid; today, it is called condensate or natural gasoline.)

DISTILLATE FUEL OILS
A term denoting products of refinery distillation sometimes referred to as middle distillates, i.e., kerosine, diesel fuel, home heating oil.

DISTILLATION
The refining process of separating crude oil components by heating and subsequent condensing of the fractions by cooling.

DISTILLATION SYSTEM
A small, temporary "refinery" (200 to 1,000 b/d) set up at a remote drilling site to make diesel fuel and low-grade gasoline from available crude oil for the drilling engines and auxiliary equipment.

DIVERTER SYSTEM
An assembly of nipples and air-actuated valves welded to a well's surface or conductor casing for venting a gas kick (q.v.) encountered in relatively shallow offshore wells. In shallow wells there is often insufficient overburden pressure around the base of the conductor casing to prevent the gas from a substantial kick from blowing out around the casing. When a kick occurs, the blowout preventer is closed and the valves of the diverter system open to vent the gas harmlessly to the atmosphere.

DIVESTITURE
Specifically as it relates to the industry, to break up, to fragment an integrated oil company into individual, separate companies, each to be permitted to operate within only a single phase of the oil business: exploration, production, transportation, refining or marketing.

DIVISION ORDER
A contract of sale to the buyer of crude oil or gas directing the buyer to pay for the product in the proportions set forth in the contract. Certain amounts of payment go to the operator of the producing property, the royalty owners, and others having an interest in the production. The purchaser prepares the division order after determining the basis of ownership and then requires that the several owners of the oil being purchased execute the division order before payment for the oil commences.

DOCTOR SWEET
A term used to describe certain petroleum products that have been treated to remove sulfur compounds and mercaptans that are the sources of unpleasant odors. A product that has been so treated is said to be "sweet to the doctor test."

DOCTOR TEST
A qualitative method of testing light fuel oils for the presence of sulfur compounds and mercaptans, substances that are potentially corrosive and impart an objectionable odor to the fuel when burned.

DOGHOUSE
A portable, one-room shelter (usually made of light tank-iron) at a well site for the convenience and protection of the drilling crew, geologist, and

others. The doghouse serves as lunch room, change house, dormitory, and for keeping small supplies and records.

DOG IT, TO
To do less than one's share of work; to hang back; to drag one's feet.

DOG LEG
A deviation in the direction of a ditch or the bore hole of a well; a sharp bend in a joint of pipe. *See* Key seat.

DOG ROBBER
A loyal aid or underling who does disagreeable or slightly unorthodox (shady) jobs for his boss; a master of the "midnight requisition."

DOLLY
Metal rollers fixed in a frame and used to support large diameter pipe as it is being turned for welding; a small, low platform with rollers or casters used for moving heavy objects.

DOME
An incursion of one underground formation into the one above it; an upthrust, as a salt dome, that penetrates overlying strata.

DONKEY PUMP
Any small pumping device used on construction jobs or other temporary operations.

DOODLE BUG
A witching device; a twig or branch of a small tree (peach is favored by some witchers) which when held by an "expert" practitioner as he walks over a plot of land is supposed to bend down locating a favorable place to drill a well; a popular term for any of the various geophysical prospecting equipment.

DOPE
Any of various viscous materials used on pipe or tubing threads as a lubricant, and to prevent corrosion; a tarbase coating for pipelines to prevent corrosion.

DOUBLE-ACTING PUMP
A reciprocating pump (plunger pump) with two sets of suction and discharge valves permitting it to pump fluid during the forward and backward movement of each plunger. (Single-action pumps discharge on the forward stroke and draw in fluid on the return stroke.)

DOUBLES
Drill pipe and tubing pulled from the well two joints at a time. The two joints

45

make a stand (q.v.) of pipe that is set back and racked in the derrick. Three-joint stands are called "thribbles," fours are "fourbles."

DOUBLING YARD
An area convenient to a large pipeline construction project where line pipe is welded together in two-joint lengths preparatory to being transported to the job and strung along the right-of-way.

DOWN-COMER
A pipe in which the flow of liquid or gas is downward.

DOWN HOLE
A term to describe tools, equipment, and instruments used in the well bore; also, conditions or techniques applying to the well bore.

DOWN IN THE BIG HOLE
A slang expression meaning to shift down into the lowest low gear.

DOWSING RODS
See Doodle bug

DOWNSTREAM
A term referring to industry operations beyond the producing phase of the business, i.e., refining and marketing.

DOZER
Bulldozer, a crawler-type tractor equipped with a hydraulically operated blade for excavating and grading.

DRAG UP, TO
To draw the wages one has coming and quit the job; an expression used in the oil fields by pipeline construction workers and temporary or day laborers.

DRAINAGE
Migration of oil or gas in a reservoir owing to a pressure reduction caused by production from wells drilled into the reservoir. Local drainage is the movement of oil and gas toward the well bore of a producing well.

DRAINAGE UNIT
The maximum area in an oil pool or field that may be drained efficiently by one well so as to produce the maximum amount of recoverable oil or gas in such an area.

DRAKE, COL. EDWIN L.
The man who drilled the country's first oil well near Titusville, Pennsylvania, in 1859 to a depth of 69½ feet using crude cable-tool equipment.

DRAWING THE FIRES
Shutting down a refinery unit in preparation for a turnaround (q.v.).

DRAW WORKS
The collective name for the hoisting drum, cable, shaft, clutches, power take off, brakes, and other machinery used on a drilling rig. Draw works are located on one side of the derrick floor, and serve as a power-control center for the hoisting gear and rotary elements of the drill column.

DRESS-UP CREW
A right-of-way gang that clean up after the construction crews have completed their work. The dress-up crew smooths the land, plants trees, grass, and builds fences and gates.

DRESSER SLEEVE
A slip-type collar that connects two lengths of plain-end (threadless) pipe. This type sleeve connection is used on small-diameter, low pressure lines.

DRIFT MANDREL
A device used to check the size of casing and tubing before it is run. The drift mandrel (jack rabbit) is put through each joint of casing and tubing to make certain the inside diameters are sizes specified for the particular job.

DRILL BIT, DRAG
A type of old-style drilling tool in which the cutting tooth or teeth were the shape of a fish tail. Drilling was accomplished by the tearing and gouging action of the bit, and was efficient in soft formations; the forerunner of the modern, three-cone roller bit.

DRILL BIT, FISH-TAIL
A drag bit. See Drill bit, Drag.

DRILL BIT, INSERT
A bit with super-hard metal lugs or cutting points inserted in the bit's cutting cones; a rock bit with cutting elements added that are harder and more durable than the teeth of a mill-tooth bit (q.v.).

DRILL BIT, MILL-TOOTH
A bit with cutting teeth integral to the metal of the cones of the bit; a non-insert bit. Mill-tooth bits are used in relatively soft formations found at shallow depths.

DRILL BIT, ROTARY
The tool attached to the lower end of the drill pipe; a heavy steel "head" equipped with various types of cutting or grinding teeth, some are fixed, some turn on bearings. A hole in the bottom of the drill permits the flow of drilling mud being pumped down through the drill pipe to wash the cuttings to the surface and also cool and lubricate the bit.

DRILL COLLAR
A heavy, tubular connector between drill pipe and bit. Originally, the drill

collar was a means of attaching the drill bit to the drill pipe and to strengthen the lower end of the drill column which is subject to extreme compression, torsion, and bending stresses. Now the drill collar is used to concentrate a heavy mass of metal near the lower end of the drill column. Drill collars were once a few feet long and weighed 400 or 500 pounds. Today, because of increased bit pressure and rapid rotation, collars are made up in 1,000-foot lengths and weigh 50-100 tons.

DRILL COLLAR, SQUARE
A type of drill collar whose cross section is square instead of circular as in a more conventional collar. Square drill collars are used to prevent or minimize the chances of becoming hung up or stuck in a dog leg down-hole. The square corners on the collar, which is located just above the drill bit in the string, act as a reamer and tend to keep the hole passable for the drill pipe.

DRILLER
One who operates a drilling rig; the person in charge of drilling operations and who supervises the drilling crew.

DRILLER'S LOG
A record kept by the driller showing the following: when the well was spudded in, the size of the hole, the bits used, when and at what depth various tools were used, the feet drilled each day, the point at which each string of casing was set, and any unusual drilling condition encountered. In present day wells, the driller's log is supplemented by electrical well logs.

DRILLING CONTRACTOR
A person or company whose business is drilling wells. Wells are drilled on a per-foot basis, others are contracted for on a day rate. *See* Turnkey job.

DRILLING FLOOR
Derrick floor; the area where the driller and his crew work.

DRILLING ISLAND
A man-made island constructed in water 10- to 50-feet deep by dredging up the lake or bay bottom to make a foundation from which to drill wells. This procedure is used for development drilling, rarely in wildcatting.

DRILLING JARS
A jointed section in a string of cable tools made with slack or play between the joints. If the bit becomes lodged in the hold, the sudden jar or impact developed by taking up of the slack in the jars aids in freeing the bit.

DRILLING LOG
See Driller's Log

DRILLING MAST
A type of derrick consisting of two parallel legs, in contrast to the conventional four-legged derrick in the form of a pyramid. The mast is held upright by guy wires. This type mast is generally used on shallow wells or for reconditioning work. An advanced type of deep-drilling rig employs a mast-like derrick of two principal members with a base as an integral part of the mast.

DRILLING MUD
A special mixture of clay, water, and chemical additives pumped downhole through the drill pipe and drill bit. The mud cools the rapidly rotating bit; lubricates the drill pipe as it turns in the well bore; carries rock cuttings to the surface; and serves as a plaster to prevent the wall of the bore hole from crumbling or collapsing. Drilling mud also provides the weight or hydrostatic head to prevent extraneous fluids from entering the well bore and to control downhole pressures that may be encountered.

DRILLING MUD DENSITY
The weight of drilling mud expressed in pounds per U.S. gallon or in pounds per cubic foot. Density of mud is important because it determines the hydrostatic pressure (q.v.) the mud will exert at any particular depth in the well. In the industry, mud weight is synonymous with mud density. To "heavy up on the mud" is to increase its density.

DRILLING PERMIT
See Well permit

DRILLING PLATFORM
An offshore structure with legs anchored to the sea bottom. The platform, built on a large-diameter pipe frame, supports the drilling of a number of wells from the location. As many as 60 wells have been drilled from one large offshore platform.

DRILLING SPOOL
The part of the drawworks that holds the drilling line; the drum of drilling cable on which is spooled the wire line that is threaded over the crown block sheaves and attached to the traveling block.

DRILLING TENDER
A barge-like vessel that acts as a supply ship for a small, offshore drilling platform. The tender carries pipe, mud, cement, spare parts, and in some instances, provides crew quarters.

DRILL PIPE
Heavy, thick-walled steel pipe used in rotary drilling to turn the drill bit and to provide a conduit for the drilling mud. Joints of drill pipe are about 30-feet long.

DRILL, ROTARY-PERCUSSION

A drill bit that rotates in a conventional manner but at the same time acts as a high-frequency pneumatic hammer, producing both a boring and a fracturing action simultaneously. The hammer-like mechanism is located just above the bit and is actuated by air, liquid, or high-frequency sound waves.

DRILL SHIP

A self-propelled vessel, a ship equipped with a derrick amidships for drilling wells in deep water. A drill ship is self-contained, carrying all of the supplies and equipment needed to drill and complete a well.

DRILL STEM

The drill pipe. In rotary drilling, the bit is attached to the drill stem or drill column which rotates to "dig" the hole.

DRILL-STEM TEST

A method of obtaining a sample of fluid from a formation using a "formation-tester tool" attached to the drill stem. The tool consists of a packer (q.v.) to isolate the section to be tested and a chamber to collect a sample of fluid. If the formation pressure is sufficient, fluid flows into the tester and up the drill pipe to the surface.

DRIP

A small in-line tank or condensing chamber in a pipeline to collect the liquids that condense out of the gas stream. Drips are installed in low places in the line and must be "blown" or emptied periodically.

DRIP GASOLINE

Natural gasoline recovered at the surface of a well as the result of the separation of certain of the liquid hydrocarbons dissolved in the gas in the formation; gasoline recovered from a drip (q.v.) in a field gas line; casing-head gasoline.

DRIP OILER
See Wick oiler

DRIVE PIPE

A metal casing driven into the bore hole of a well to prevent caving of the walls and to shut off surface water. The drive pipe, first used in an oil well by Colonel Drake, was the forerunner of the modern conductor or surface pipe. *See* Casing.

DRIVE THE HOOPS, TO

To tighten the staves of a wooden stock tank by driving the metal bands or hoops down evenly around the circumference of the tank. Early-day lease tanks were made of redwood in the shape of a truncated cone (nearly cylindrical). Metal bands like those on a wooden barrel held the staves together. Once a year or so, the hoops had to be driven to tighten the seams between the staves to prevent leaks. Today wooden tanks are used on leases to handle salt water and other corrosive liquids. Their staves are held together with steel rods equipped with turnbuckles for keeping the tank water-tight.

DRUM
A 55-gallon metal barrel; a standard container used for shipping lubricating oil and other petroleum products.

DRY GAS
Natural gas from the well free of liquid hydrocarbons; gas that has been treated to remove all liquids; pipeline gas.

DRY HOLE
An unsuccessful well; a well drilled to a certain depth without finding oil; a "duster" (q.v.).

DRY-HOLE MONEY
Money paid by one or more interested parties (those owning land or a lease nearby) to an operator who drills a well that is a dry hole. The well whether successful or dry serves to "prove their land," providing useful information. Before the well is drilled, the operator solicits dry hole "contributions" and in return for financial assistance agrees to furnish certain information to the contributors.

DRY-HOLE PLUG
A plug inserted in a well that is dry to seal off the formations that were penetrated by the bore hole. This treatment prevents salt water, often encountered in "dry holes," from contaminating other formations. See Plugging a well.

D.S.T.
Drill stem test (q.v.).

DUAL COMPLETION
The completion of a well in two separate producing formations, each at different depths. Wells sometimes are completed in three or even four separate formations with four strings of tubing inserted in the casing. This is accomplished with packers (q.v.) that seal off all formations except the one to be produced by a particular string of tubing.

DUAL DISCOVERY
A well drilled into two commercial pay zones, two separate producing formations, each at a different depth.

DUAL-FUEL ENGINES
Engines equipped to run on liquid as well as gaseous fuel. Stationary engines in the field have modifications made to their carburetors that

permit them to operate either on gasoline or natural gas. In some installations, when the gasoline supply is used up, the engine is switched to natural gas automatically.

DUBAI STORAGE TANKS
A specially designed, under-water storage tank the shape of an inverted funnel, built by Chicago Bridge & Iron for Dubai Petroleum Company. The tanks have no bottoms and rest on the sea floor supported on their rims. Oil from fields on shore is pumped into the top of the tanks under pressure forcing the sea water out the bottom. The offshore tanks which are more than 100 feet tall also serve as single-point moorings for tankers taking on crude.

DUBBS, CARBON PETROLEUM
Mr. Dubbs, a petroleum chemist, developed a cracking process that found wide acceptance in the 1920s, and was almost as popular as the Burton still which was developed earlier for Standard Oil Company of Indiana.

DUCK'S NEST
Colloquial term for a standby drilling mud tank or pit used to hold extra mud, or as an overflow in the event of a gas "kick" (q.v.).

DUMP BOX
A heavy wooden or metal box where the contents of a cable-tool well's bailer is emptied. The end of the bailer is lowered into the box which pushes the dart up upwards, unseating the ball valve and permitting the water, mud, and rock cuttings to empty into the box and slush-pit launder (q.v.).

DUMP FLOODING
An unusual secondary recovery technique that uses water from a shallow water bed above the producing pay to flood the oil producing interval. The water from the aquifer (q.v.) enters the injection string by its own pressure. The weight of the hydrostatic column (water column) produces the necessary force for it to penetrate the oil formation, pushing the oil ahead of it to the producing wells in the field.

DUSTER
A completely dry hole; a well that encounters neither gas nor liquid at total depth.

DUTCHMAN
The threaded portion of a length of pipe or nipple twisted or broken off inside a collar or other threaded fitting. Threads thus "lost" in a fitting have to be cut out with a chisel or cutting torch.

DWT
Deadweight-ton; a designation for the size or displacement of a ship, e.g. 100,000 dwt crude oil tanker.

DYNAMIC STATIONING
A method of keeping a drill ship or semi-submersible drilling platform on target, over the hole during drilling operations where the water is too deep for the use of anchors. This is accomplished by the use of thrusters (q.v.) activated by underwater sensing devices that signal when the vessel has moved a few degrees off its drilling station.

E

EARNEST MONEY
A sum of money paid to bind a financial transaction prior to the signing of a contract; hand money.

EASY DIGGING
A soft job; an assignment of work that can be handled without much exertion.

ECONOMIC DEPLETION
The reduction in the value of a wasting asset (q.v.) by removing or producing the minerals.

EFFLUENT
The discharge or out-flow from a manufacturing or processing plant; outfall; drainage.

ELECTRIC (WELL) LOG
An electrical survey of a well's bore hole before it is cased, which reflects the degree of resistance of the rock strata to electric current. From the results of the survey, geologists are able to determine the nature of the rock penetrated by the drill and some indication of its permeability.

ELECTRIC WELL-LOGGING
The procedure of lowering electrical instruments into the well bore to test the density and other characteristics of rock formations penetrated.

ELECTROMOTIVE FORCE
The amount of energy derived from an electrical source per unit quantity of electricity passing through the source. *See* Thermocouple.

ELEVATORS
A heavy, hinged clamp attached to the hook and traveling block by bail-like arms, and used for lifting drill pipe, casing, and tubing and lowering them into the hole. In hoisting a joint of drill pipe, the elevators are latched on to the pipe just below the tool joint, (coupling) which prevents the pipe from slipping through the elevators.

E.M.F.
Electromotive force (q.v.)

EMINENT DOMAIN
The right of a government body or public utility (common carrier) to take private property for public use by condemnation proceedings (q.v.).

END-O
The command given by one worker to another or to a group to lift together and move an object forward; a signal to "put out" in a big lift.

END POINT
The point indicating the end of an operation or the point at which a certain definite change is observed. In the analysis of liquids such as gasoline, the end point is the temperature at which the liquid ceases to distill over. End points are of value in predicting certain performance characteristics of gasoline.

END PRODUCTS
Material, substances, goods for consumer use; finished products.

END USE
Ultimate use; consumption of a product by a commercial or industrial customer.

ENERGY SOURCES
Petroleum — coal — hydropower — nuclear — geothermal (q.v.) — synthetic fuels (q.v.) — tides — solar.

ENHANCED RECOVERY
Sophisticated recovery methods for crude oil which go beyond the more conventional secondary recovery techniques of pressure maintenance and water flooding. Enhanced recovery methods now being used include micellar-surfactant (q.v.), steam drive, polymer, miscible hydrocarbon, CO_2, and steam soak. ER methods are not restricted to secondary or even tertiary projects. Some fields require the application of one of the above methods even for initial recovery of crude oil.

ENRICHED-GAS INJECTION
A secondary recovery method involving the injection of gas rich in inter-mediate hydrocarbons or enriched by addition of propane, butane, or pentane on the surface or in the well bore as the gas is injected.

ENTRAINED OIL
Oil occurring as part of the gas stream, but as a relatively small percen-tage of total flow. Special separators are used to remove the liquid from the gas stream.

ENTRY POSITION
A starting job with a company usually sought by a young man or woman just out of school who wishes to get into the business at whatever level—with the expectation of becoming president in due time.

EPA
Environmental Protection Agency, U.S. Department of the Interior

E.P. LUBRICANTS
Extreme pressure lubricating oils and greases which contain substances added to prevent metal to metal contact in highly loaded gears and turntables.

EQUITY CRUDE
In cases where a concession is owned jointly by a host government and an oil company, the crude produced which belongs to the oil company is known as equity crude, as opposed to buy-back (participation crude). The cost of equity crude is calculated according to the posted price. See Buy-back.

ER
Enhanced recovery of crude oil (q.v.).

ESCAPE BOOMS
Devices used on offshore drilling or production platforms for emergency escape of personnel in the event of a fire or explosion. They consist of counter-weighted arms supporting a buoyant head. When the arms are snapped loose from the platform, they fall outward, the head descending to the water. The workers then slide down a lifeline to the floating head.

ETHANE
A simple hydrocarbon associated with petroleum. Ethane is a gas at ordinary atmospheric conditions.

EVAPORATION PIT
An excavation dug to contain oil field salt water or brine which is disposed of by evaporation. Great amounts of salt water are produced with crude oil in some oil fields, particularly older fields.

EXPANSION FIT
See Shrink fit

EXPANSION JOINT
A section of piping constructed in such a way as to allow for expansion and contraction of the pipe connections without damaging the joints. Specially fabricated, accordion-like fittings are used as expansion joints in certain in-plant hookups where there are severe temperature changes.

EXPANSION LOOP
A circular loop (360° bend) put in a pipeline to absorb expansion and contraction caused by heating and cooling, without exerting a strain on pipe or valve connections.

EXPANSION-ROOF TANK
A storage or working tank (q.v.) with a roof made like a slip joint. As the vapor above the crude oil or volatile product expands with the heat of the day, the roof-and-apron section of the tank moves upward permitting the gas to expand without any loss to the atmosphere. The telescoping roof, as it moves up and down, maintains a gas-tight seal with the inner wall of the tank.

EXPLORATION ACTIVITIES
The search for oil and gas. Exploration activities include aerial surveying, geological studies, geophysical surveying, coring, and drilling of wildcat wells.

EXPLORATORY WELLS
Wells drilled to find the limits of an oil-bearing formation, often referred to as a pool, only partly developed. *See* Step-out well.

EXPLOSION-PROOF MOTORS
A totally enclosed electric motor with no outside air in contact with the motor windings; an enclosed brushless motor. Cooling is by conduction through the frame and housing.

EXTENSION TEST
See Outpost well

EXTERNAL CASING PACKER
A device used on the outside of a well's casing to seal off formations or to protect certain zones. Often used downhole in conjunction with cementing. The packer is run on the casing and when at the proper depth, it may be expanded against the wall of the bore hole hydraulically or by fluid pressure from the well.

EYEBALL
To align pipe connections or a temporary construction with the eye only; to inspect carefully.

F

FAIL SAFE
Said of equipment or a system so constructed that, in the event of failure or malfunction of any part of the system, devices are automatically activated to stabilize or secure the safety of the operation.

FAIRLEAD
A guide for ropes or lines on a ship to prevent chaffing; a sheave supported by a bracket protruding from the cellar deck of a semisubmersible drilling platform over which an anchor cable runs. Some large floating platforms have anchor lines made up of lengths of chain and cable.

FAIRWAY
A shipping lane established by the U.S. Coast Guard in Federal offshore waters. Permanent structures such as drilling and production platforms are prohibited in a fairway which significantly curtails oil activity in some offshore areas.

FANNING THE BOTTOM (OF THE BORE HOLE)
Drilling with very little weight on the drill bit in the hope of preventing the bit from drifting from the vertical and drilling a crooked hole. Fanning the bottom, however, is considered detrimental to the drill string, by some authorities, as reduced weight on the bit causes more tension on the drill pipe, resulting in pipe and collar fatigue.

FARM BOSS
A foreman who supervises the operations of one or more oil-producing leases.

FARM IN
An arrangement whereby one oil operator "buys in" or acquires an interest in a lease or concession owned by another operator on which oil or gas has been discovered or is being produced. Often farm-ins are negotiated to assist the original owner with development costs and to secure for the buyer a source of crude or natural gas. See Farm-out agreement.

FARM OUT
The name applied to a leasehold held under a farm-out agreement (q.v.).

FARM OUT AGREEMENT
A form of agreement between oil operators whereby the owner of a lease who is not interested in drilling at the time agrees to assign the lease or a portion of it to another operator who wishes to drill the acreage. The assignor may or may not retain an interest (royalty or production payment) in the production.

FARMER'S OIL
An expression meaning the land owner's share of the oil from a well drilled on his property; royalty, usually one-eighth of the produced oil free of any expense to the landowner.

FARMER'S SAND
A colloquial term for "the elusive oil-bearing stratum which many land-owners believe lies beneath their land, regardless of the results of exploratory wells."

FAST LINE
On a drilling rig, the fast line is the cable spooled off or on the hoisting drum of the draw works; the line from the hoisting drum that runs up through the derrick to the first sheave in the crown block.

FAT OIL
The absorbent oil enriched by gasoline fractions in an absorption plant. After absorbing the gasoline fractions, the gasoline is removed by distillation, leaving the oil "lean" and ready for further use to absorb more gasoline fractions from the natural gas stream.

FAULT
A fracture in the earth's crust accompanied by a shifting of one side of the fracture with respect to the other side; the point at which a geological strata "breaks off" or is sheared off by the dropping of a section of the strata by settling.

FEA
Federal Energy Agency (q.v.).

FEDERAL ENERGY AGENCY
The government agency that administers the Federal Energy Law whose regulations and directives govern the activities of the oil and gas industry.

FEDERAL LEASE
An oil or gas lease on Federal land issued under the Act of February 25, 1920, and subsequent legislation.

FEE
The title or ownership of land; short for "owned-in-fee." The owner of the fee holds title to the land.

FEE ROYALTY
The lessor's share of oil and gas production; landowner's royalty.

FEE SIMPLE
Land or an estate held by a person in his own right without restrictions.

FEED OR FEEDSTOCK
Crude oil or other hydrocarbons that are the basic materials for a refining or manufacturing process.

FEEDER LINE
A pipeline; a gathering line tied into a trunk line.

FEMALE CONNECTION
A pipe, rod, or tubing coupling with the threads on the inside.

FIELD
The area encompassing a group of producing oil and gas wells; a pool.

FIELD COMPRESSION TEST
A test to determine the gasoline content of casinghead or wet gas.

FIELD POTENTIAL
The producing capacity of a field during a 24-hour period.

FIELD TANKS
Stock tanks (q.v.).

FILTER CAKE
A plaster-like coating of the bore hole resulting from the solids in the drilling fluid adhering and building up on the wall of the hole. The buildup of "cake" can cause serious drilling problems including the sticking of the drill pipe. *See* Differential-pressure sticking.

FILTRATE
The solid material in drilling mud. When filtrate is deposited on the wall of the bore hole of the well forming a thick, restrictive layer, it is referred to as filter cake.

FILTRATION-LOSS QUALITY OF MUD
A drilling-mud quality measured by putting a water-base mud through a filter cell. The mud-solids deposited on the filter is filter cake (q.v.) and is a measure of the water-loss quality of the drilling mud. Mud mixtures with low water-loss are desirable for most drilling operations.

FINES
Minute particles of a solid substance—rock, coal, or catalytic material—too small to be used or handled efficiently, and that must be removed by screening.

FINGER BOARD
A rack high in the derrick made to support, in orderly fashion, the upper ends of the tubing stands that are pulled from the well and set back (q.v.) in the derrick.

FINGER PIER
A jetty or bridge-type structure extending from the shore out into a body of water to permit access to tankers and other vessels where water depth is not sufficient to allow docking at the shore.

FINGERING
Rivulet-like infiltration of water or gas into an oil-bearing formation as a result of failure to maintain reservoir pressure, or as the result of taking oil in excess of maximum efficiency rates (MER) (q.v.). *See* Channel.

FIRE FLOODING
See In-situ combustion

FIRE-WATER SUPPLY
A pond or tank containing water used exclusively for fire fighting.

FIRING LINE
In pipeline construction, the part of the project where the welding is being done.

FIREWALL
An earthen dike built around an oil tank to contain the oil in the event the tank ruptures or catches fire.

FISH, THE
Anything lost down the drill hole; the object being sought downhole by the fishing tools.

FISHING JOB
The effort to recover tools, cable, pipe, or other objects from the well bore which may have become detached while in the well or been dropped accidentally into the hole. Many special and ingeniously designed fishing tools are used to recover objects lost downhole.

FISHING TOOLS
Special instruments used in recovering objects lost in a well. Although there are scores of standard tools used in fishing jobs, some are specially designed to retrieve particular objects.

FITTINGS
Small pipes (nipples), couplings, elbows, unions, tees, swages used to make up a system of piping.

FIVE-SPOT WATER FLOOD PROGRAM
A secondary recovery operation where four input or injection wells are located in a square pattern with the production well in the center, a layout similar to a five-of-spades playing card. The water from the four injection wells moves through the formation flooding the oil toward the production well.

FLAG THE LINE
To tie pieces of cloth on the swab line at measured intervals to be able to tell how much line is in the hole when coming out with the line.

FLAMBEAU LIGHT
A torch used in the field for the disposal of casinghead gas produced with oil when the gas is without a market or is of such small quantity as to make it impractical to gather for commercial use. The use of flambeau lights is now regulated under state conservation laws.

FLAME ARRESTER
A safety device installed on a vent line of a storage or stock tank that, in the event of lightning or other ignition of the venting vapor, will prevent the flame from flashing to the vapors inside the tank.

FLAME-JET DRILLING
A drilling technique that uses rocket fuel to burn a hole through rock strata. This leaves a ceramic-like sheath on the walls of the bore hole eliminating the need for casing.

FLAME SNUFFER
An attachment to a tank's vent line that can be manually operated to snuff out a flame at the mouth of the vent line; a metal clapper-like valve that may be closed by pulling on an attached line.

FLAMMABLE
Term describing material which can be easily ignited. Petroleum products with a flash point (q.v.) of 80°F or lower are classed as flammable.

FLANGE
(1) A type of pipe coupling made in two halves. Each half is screwed or welded to a length of pipe and the two halves are then bolted together joining the two lengths of pipe. (2) A rim extending out from an object to provide strength or for attaching another object.

FLANGE UP
To finish a job; to bring to completion.

FLARE
(1) To burn unwanted gas through a pipe or stack. (Under conservation laws, the flaring of natural gas is illegal.) (2) The flame from a flare; the pipe or the stack itself.

FLASH OFF
To vaporize from heated charge stock; to distill.

FLASH POINT
The temperature at which a given substance will ignite.

FLEXIBLE COUPLING
A connecting link between two shafts that allows for a certain amount of misalignment between the driving and driven shaft without damage to bearings. Flexible couplings dampen vibration and provide a way to make quick hook-ups of engines and pumps which is useful in field operations.

FLOAT
(1) A long, flat-bed trailer the front end of which rests on a truck, the rear end on two dual-wheel axles. Floats are used in the oil fields for transporting long, heavy equipment. (2) The buoyant element of a fluid-level shut off or control apparatus. An air-tight canister or sphere that floats on liquids and is attached to an arm that moves up and down, actuating other devices as the liquid level rises and falls.

FLOATER
(1) A barge-like drilling platform used in relatively shallow, offshore work. (2) Any offshore drilling platform without a fixed base, e.g. semi-submersibles—drill ships—drill barges.

FLOATING ROOF TANK
A storage tank with a flat roof that floats on the surface of the oil thus reducing evaporation to a minimum. The roof rests on a series of pontoons whose buoyancy supports the roof proper; a floater.

FLOATING THE CASING
A method of lowering casing into very deep bore holes when there exists the danger of the casing joints separating because of the extreme weight or tension on the upper joints. In floating, the hole is filled with fluid and the casing is plugged before being lowered into the hole. The buoyant effect of the hollow column of casing displacing the fluid reduces the weight and the tension on the upper joints. When the casing is in place, the plug is drilled out.

FLOAT VALVE
A valve whose stem is actuated by an arm attached to a float; an automatic valve operated, through linkage to a float mechanism, by the change in liquid level in a tank or other vessel.

FLOODING
The use of water injected into a production formation or reservoir to increase oil recovery. *See* Secondary recovery.

FLOOR MEN
Members of the drilling crew (usually two) who work on the derrick floor.

FLOW BEAN
A drilled plug in the flow line at the well-head that controls the oil flow to a desired rate. Flow beans are drilled with different sized holes for different flow rates.

FLOW CHART
A replaceable, paper chart on which flow rates are recorded by an actuated arm and pen. *Also* Pressure chart; Temperature chart.

FLOW SHEET
A diagramatic drawing showing the sequence of refining or manufacturing operations in a plant.

FLOW STRING
The string of casing or tubing through which oil from a well flows to the surface. *Also* Oil string; pay string; capital string; production string.

FLOW TANK
A lease tank into which produced oil is run after having gas or water removed; production tank.

FLOW TREATER
A single unit that acts as an oil and gas separator, an oil heater, and an oil and water treater.

FLUID CATALYTIC CRACKING UNIT
A large refinery vessel for processing reduced crude, naphthas, or other intermediates in the presence of a catalyst. Catalytic cracking is regarded as the successor to thermal cracking as it produces less gas and highly volatile material; it provides a motor spirit of 10-15 octane numbers higher than that of thermally-cracked product. The process is also more effective in producing *iso*-paraffins and aromatics which are of high anti-knock value.

FLUID END (OF A PUMP)
The end of the pump body where the valves (suction and discharge) and the pump cylinders are located. The fluid end of a reciprocating pump is accessible by removing the cylinder heads which exposes the pistons or pump plungers. The cylinders or liners in most pumps are removable and can be changed for others with larger or smaller internal diameters. Inserting smaller liners and pistons permits pumping at higher pressure but at a reduced volume.

FLUIDICS
Pertains to the use of fluids (and air) in instrumentation. Fluidics is defined as "engineering science pertaining to the use of fluid-dynamic phenomena to sense, control, process information, and actuate." Fluidics provide a

reliable system far less expensive than explosion-proof installations required with electrical instrumentation on offshore rigs.

FLUID LEVEL
The distance between the wellhead and the point to which the fluid rises in the well.

FLUSH PRODUCTION
The high rate of flow of a good well immediately after it is brought in.

FLUSHING OILS
Oils or compounds formulated for the purpose of removing used oil, decomposed matter, metal cuttings, and sludge from lubricating passages and engine parts.

FLUXING
To soften a substance with heat so that it will flow; to lower a substance's fusing point.

FOOT-POUND
A unit of energy or work equal to the work done in raising one pound the height of one foot against the force of gravity.

FOOT VALVE
A type of check valve (q.v.) used on the "foot" or lower end of a suction-pipe riser to maintain the column of liquid in the riser when the liquid is being drawn upward by a pump.

FORCED-DRAFT BURNER
Crude oil disposal equipment on offshore platforms. The burner, mounted on a boom or an extension of the deck, burns crude oil during testing operations. Gas, air, and water manifolded with the test-crude stream result in complete combustion of the oil.

FOREIGN TAX CREDIT
Taxes paid a foreign government by a U.S. company on its overseas oil operations that are creditable against taxes owed the U.S. government. Production sharing by a U.S. company with a foreign government or one of its agencies represents oil royalty payments, not taxes creditable in the U.S., according to the Internal Revenue Service.

FOREIGN TRADE ZONE
An area in the United States where imported oil, reduced crude, or intermediates are processed.

FORMATION
A sedimentary bed or series of beds sufficiently alike or distinctive to form an identifiable geological unit.

FORMULATION
The product of a formula, i.e., a plastic, blended oils, gasolines; any material with two or more components or ingredients.

FOSSIL ENERGY
Energy derived from crude oil, natural gas, and coal.

FOSSIL FUEL
See Fossil energy

FOUR-CYCLE ENGINE
An internal combustion engine in which the piston completes four strokes—intake, compression, power, and exhaust—for each complete cycle. The Otto-cycle engine; four-stroke cycle engine.

FOURBLE
A stand (q.v.) of drill pipe or tubing consisting of four joints. In pulling pipe from the well, every fourth joint is unscrewed and the four-joint stand is set back and racked in the derrick. This is not a common practice; the usual stands are of two and three joints.

FPC
Federal Power Commission, an agency of the Federal government; a regulatory body having to do with oil and gas matters such as pricing and trade practices.

FRAC JOB
See Hydraulic fracturing

FRACTION
A separate, identifiable part of crude oil; the product of a refining or distillation process.

FRACTIONATOR
A tall, cylindrical refining vessel where liquid feedstocks are separated into various components or fractions.

FREEZE BOX
An enclosure for a water-pipe riser that is exposed to the weather. The freeze box or frost box surrounding the pipe is filled with sawdust, manure or other insulating material.

FREON
A trademark applied to a group of halogenated hydrocarbons having one or more florine atoms in the molecule; a refrigerant.

FRESNO SLIP
A type of horse-drawn, earth-moving or cutting scoop with curved runners

or supports on the sides and a single long handle used to guide the scoop blade into the earth or material being moved.

FROST UP
Icing of pipes and flow equipment at the wellhead of a high-pressure gas well. The cooling effect of the expanding gas, as pressure is reduced, causes moisture in the atmosphere to condense and freeze on the pipes.

FUEL-AIR RATIO
The ratio of fuel to air by weight in an explosive mixture which is controlled by the carburetor in an internal combustion engine.

FUEL OIL
Any liquid or liquefiable petroleum product burned for the generation of heat in a furnace; or for the generation of power in an engine, exclusive of oils with a flash point below 100°F.

FULL BORE
Designation for a valve, ram or other fitting whose opening is as large in cross section as the pipe, casing or tubing it is mounted on.

FULLER'S EARTH
A fine, clay-like substance used in certain types of oil filters.

FULLY INTEGRATED
Said of a company engaged in all phases of the oil business, i.e., production, transportation, refining, marketing. *See* Integrated oil company.

FUNGIBLE
Products that are or can be commingled for the purpose of being moved by product pipeline. Interchangeable.

FURFURAL
An extractive solvent of extremely pungent odor, used extensively for refining a wide range of lubricating oils and diesel fuels; a liquid aldehyde.

FUSIBLE PLUG
A fail-safe device; a plug in a service line equipped with a seal that will melt at a predetermined temperature releasing pressure that actuates shut-down devices; a meltable plug.

G

GAMMA RAY — GAMMA RAY LOGGING
A well-logging technique wherein a well's bore hole is bombarded with gamma rays from a gamma ray emitting device to induce output signals that are then recorded and transmitted to the surface. The gamma ray signals thus picked up indicate to the geologist the relative density of the rock formation penetrated by the well bore at different levels.

GAMMA RAY LOGGING
See Natural gamma ray logging

GAMMA RAYS
Minute quantities of radiation emitted by substances that are radioactive. Subsurface rock formations emit radiation quantum that can be detected by well-logging devices, and which indicate the relative densities of the surrounding rock.

GANG PUSHER
A pipeline foreman; the man who runs a pipeline or a connection gang; a pusher.

GANG TRUCK
A light or medium-size, flat-bed truck carrying a portable dog house or man rack where the pipeline repair crew rides to and from the job. The pipeliners' tools are carried in compartments beneath the bed of the truck.

GAS
"Any fluid, combustible or non-combustible, which is produced in a natural state from the earth and which maintains a gaseous or rarified state at ordinary temperature and pressure conditions." Code of Federal Regulations, Title 30, Mineral Resources, Chap. II, Geological Survey, 221.2.

GAS ANCHOR
A device for the bottom-hole separation of oil and gas in a pumping well. The gas anchor (a length of tubing about 5-feet long) is inside a larger pipe with perforations at the upper end. Oil in the annulus between the well's casing and tubing enters through the perforations and is picked up by the pump; the gas goes out through the casing to the wellhead.

GAS BOTTLES
The cylindrical containers of oxygen and acetylene used in oxyacetylene welding. Oxygen bottles are tall and slender with a tapered top; acetylene bottles are shorter and somewhat larger in diameter.

GAS BUSTER
A drilling-mud/gas separator; a surge chamber on the mud-flow line where entrained gas breaks out and is vented to a flare line; the gas-free mud is returned to the mud tanks.

GAS CAP
The portion of an oil-producing reservoir occupied by free gas; gas in a free state above an oil zone.

GAS-CAP ALLOWABLE
A production allowable granted an operator who shuts in a well producing from a gas cap of an oil-producing reservoir. The allowable is transferable to another lease in the same field. The shutting in of the gas-cap producer preserves the reservoir pressure which is essential to good production practice.

GAS-CAP DRIVE
The energy derived from the expansion of gas in a free state above the oil zone which is used in the production of oil. Wells drilled into the oil zone cause a release of pressure which allows the compressed gas in the cap to expand and move downward forcing the oil into the well bores of the producing wells.

GAS-CAP FIELD
A gas-expansion reservoir in which some of the gas occurs as free gas rather than in solution. The free gas will occupy the highest portion of reservoir. When wells are drilled to lower points on the structure, the gas will expand forcing the oil down-dip and into the well bores.

GAS CONDENSATE
Liquid hydrocarbons present in casinghead gas that condense upon being brought to the surface; formerly, distillate, now condensate. Also casinghead gasoline; white oil.

GAS-CUT MUD
Drilling mud aerated or charged with gas from formations downhole. The gas forms bubbles in the drilling fluid seriously affecting drilling operations, sometimes causing loss of circulation.

GAS DISTILLATE
See Distillate

GAS ENGINE
A two or four-cycle internal combustion engine that runs on natural gas; a stationary, field engine.

GASIFICATION
Converting a solid or a liquid to a gas; converting a solid hydrocarbon

such as coal or oil shale to commercial gas; the manufacture of synthetic gas from other hydrocarbons. *See* Synthetic gas.

GAS INJECTION
Natural gas injected under high pressure into a producing reservoir through an input or injection well as part of a pressure maintenance, secondary recovery, or recycling operation.

GAS INJECTION WELL
A well through which gas under high pressure is injected into a producing formation to maintain reservoir pressure.

GASKET
Thin, fibrous material used to make the union of two metal parts pressure-tight. Ready-made gaskets are often sheathed in very thin, soft metal, or they may be made exclusively of metal, or of specially formulated rubber.

GAS KICK
See Kick

GAS LIFT
A method of lifting oil from the bottom of a well to the surface by the use of compressed gas. The gas is pumped into the hole and at the lower end of the tubing it becomes a part of the fluid in the well. As the gas expands, it lifts the oil to the surface.

GAS-LIFT GAS
Natural gas used in a gas-lift program of oil production. Lift-gas is usually first stripped of liquid hydrocarbons before it is injected into the well. And because it is a "working gas" as opposed to commercial gas, its cost per thousand cubic feet (MFC) is considerably less. Gas-lift and commercial gas commingle when produced, so when the combined gas is stripped of petroleum liquids only the formation gas is credited with the recovered liquids. This is necessary for oil and gas royalty purposes.

GAS LIQUIDS
See LP-gas

GAS LOCK
A condition that can exist in an oil pipeline when elevated sections of the line are filled with gas. The gas, because of its compressibility and penchant for collecting in high places in the line, effectively blocks the gravity flow of oil. Gas lock can also occur in suction chambers of recip-rocating pumps. The gas prevents the oil from flowing into chambers and must be vented or bled off.

GAS OIL
A refined fraction of crude oil somewhat heavier than kerosine, often used as diesel fuel.

GAS-OIL RATIO
The number of cubic feet of natural gas produced with a barrel of oil.

GASOLINE PLANT
A compressor plant where natural gas is stripped of the liquid hydrocarbons usually present in wellhead gas.

GASOLINE, RAW
The untreated gasoline-cut from the distillation of crude oil.

GASOLINE, STRAIGHT-RUN
The gasoline-range fraction distilled from crude oil. Virgin naphtha.

GASSER
A commercial, natural-gas well

GAS TURBINE
An engine in which vapor (other than steam) is directed against a series of turbine blades. The energy contained in the rapidly expanding vapors or gases turns the rotors to produce the engine's power.

GAS WELDING
Welding with oxygen and acetylene or with oxygen and another gas. *See* Oxyacetylene welding.

GATE
Short for gate valve; common term for all pipeline valves.

GATE VALVE
A pipeline valve made with a wedge-shaped disk or "tongue" that is moved from open to closed position (up to down) by the action of the threaded valve stem.

GATHERING FACILITIES
Pipelines and pumping units used to bring oil or gas from production leases by separate lines to a central point, i.e., a tank farm, or a trunk pipeline.

GATHERING SYSTEM
See Gathering facilities

GAUGE HATCH
An opening in the roof of a stock or storage tank, fitted with a hinged lid, through which the tank may be gauged and oil samples taken. *See* Hatch.

GAUGE HOLE
A gauge hatch (q.v.).

GAUGE LINE
A reel of steel measuring tape, with a bob attached, held in a frame

equipped with handle and winding crank used in gauging the liquid level in tanks. To prevent striking sparks, the bob is made of brass or other non-sparking material or sheathed in a durable plastic. The tip of the bob is point zero on the gauge column.

GAUGER, FIELD
A person who measures the oil in a stock or lease tank, records the temperature, checks the sediment content, makes out a run ticket, and turns the oil into the pipeline. A gauger is the pipeline company's agent and, in effect, "buys" the tank of oil for his company.

GAUGE TANK
A tank in which the production from a well or a lease is measured.

GAUGE TAPE
Gauge line (q.v.).

GEARBOX
The enclosure or case containing a gear train or assembly of reduction gears; the case containing a pump's pinion and ring gears.

GEOCHEMISTRY
The science of chemistry applied to oil and gas exploration. By analyzing the contents of subsurface water for the presence of organic matter associated with oil deposits, geochemistry has proved to be an important adjunct to geology and geophysics in exploratory work.

GEOCHEMICAL PROSPECTING
Exploratory methods that involve the chemical analysis of rocks.

GEODESY
The branch of science concerned with the determination of the size and shape of the earth and the precise location of points on the earth's surface.

GEOLOGIC COLUMN
The vertical range of sedimentary rock from the basement rock (q.v.) to the surface.

GEOLOGICAL STRUCTURE
Layers of sedimentary rocks which have been displaced from their normal horizontal position by the forces of nature.

GEOLOGIST
A person trained in the study of the earth's crust. A petroleum geologist, in contrast to a hard-rock geologist, is primarily concerned with sedimentary rocks where most of the world's oil has been found. In general, the work of a petroleum geologist consists in searching for structural traps favorable to the accumulation of oil and gas. In addition to deciding on location to

drill, he may supervise the drilling, particularly with regard to coring, logging, and running tests.

GEOLOGRAPH
A device on a drilling rig to record the drilling rate or rate of penetration during each 8-hour tour.

GEOLOGY
The science that treats of the history of the earth and its life as recorded in the rocks.

GEOMETRY OF A RESERVOIR
A phrase used by petroleum and reservoir engineers meaning the shape of a reservoir of oil or gas.

GEOPHONES
Sensitive sound detecting instruments used in conducting seismic surveys. A series of geophones are placed on the ground at intervals to detect and transmit, to an amplifier-recording system, the reflected sound waves created by explosions set off in the course of seismic exploration work.

GEOPHYSICAL CAMP
Temporary headquarters established in the field for geophysical teams working the area. In addition to providing living quarters and a store of supplies, the camp has facilities for processing geophysical data gathered on the field trips.

GEOPHYSICAL TEAM
A group of specialists working together to gather geophysical data. Their work consists of drilling shot holes, placing explosive charges, setting out or stringing geophones, detonating shot charges, and reading and interpreting the results of the seismic shocks set off by the explosive charges.

GEOPHYSICS
The application of certain familiar physical principles—magnetic attraction, gravitational pull, speed of sound waves, the behavior of electric currents—to the science of geology.

GEOTHERMAL POWER GENERATION
The use of underground, natural heat sources, i.e., superheated water from deep in the earth, to generate steam to power turbo-electric generators.

GILSONITE
A solid hydrocarbon, with the general appearance of coal, from which gasoline, fuel oil, and coke may be obtained by a special manufacturing process.

GIN POLE
(1) An A-frame made of sections of pipe mounted on the rear of a truck bed that is used as a support or fixed point for the truck's winch line when lifting or hoisting. (2) A vertical frame on the top of the derrick, spanning the crown block, providing a support for hoisting. (3) A mast (q.v.).

GIRBITOL PROCESS
A process used to "sweeten" sour gas by removing the hydrogensulfide (H_2S).

GIRT
One of the braces between the legs of a derrick; a supporting member.

GYLCOL DEHYDRATOR
A facility for removing minute particles of water from natural gas not removed by the separator.

G.M.P.
Gallons of gasoline per 1,000 cubic feet of natural gas produced.

GO-DEVIL
A pipeline scraper; a cylindrical, plug-like device equipped with scraper blades, rollers, and wire brushes used to clean the inside of a pipeline of accumulations of wax, sand, rust, and water. When inserted in the line, the go-devil is pushed along by the oil pressure.

GONE TO WATER
A well in which the production of oil has decreased and the production of water has increased to the point where the well is no longer profitable to operate.

GOOSENECK
A nipple in the shape of an inverted U attached to the top of the swivel (q.v.) and to which the mud hose is attached.

GOOSING GRASS
Cutting grass and weeds around the lease or tank farm; shaving the grass off the ground with a sharp hoe-like tool, leaving the ground clean.

GOR
Gas/oil ratio (q.v.).

GPM
Gallons per minute

GRABLE OIL WELL PUMP
A patented, drum-and-cable pumping unit that can be installed in a wellhead cellar. The unit raises and lowers the pumping rods by winding and unwinding cable on a drum or spool. The low profile of the pumping

unit makes it ideal for use in populated areas, and to protect the beauty of the landscape.

GRADIENTS (TEMPERATURE AND PRESSURE)
The rates of increase or decrease of temperature or pressure are defined as gradients; the rate of regular or graded ascent or descent.

GRANNY HOLE
The lowest, most powerful gear on a truck.

GRANNY KNOT
A knot tied in such a way as to defy untying; an improperly tied square knot; a hatchet knot.

GRASS ROOTS
Said of a refinery or other installation built from the ground up as contrasted to a plant merely enlarged or modernized.

GRAVEL PACKING
Using gravel to fill the cavity created around a well bore as it passes through the producing zone, to prevent caving or the incursion of sand, and to facilitate the flow of oil into the well bore.

GRAVEYARD SHIFT
A tour of work beginning at midnight and ending at 8 a.m. In pipeline operations, the graveyard shift is customarily from 11 p.m. to 7 a.m. Hoot-owl shift.

GRAVIMETER
A geophysical instrument used to measure the minute variations in the earth's gravitational pull at different locations. To the geophysicist, these variations indicate certain facts about subsurface formations.

GRAVITY
(1) The attraction of the earth's mass for bodies or objects at or near the surface. (2) Short for specific gravity; API gravity. (3) To flow through a pipeline without the aid of a pump; to be pulled by the force of gravity.

GRAVITY LINE
A pipeline that carries oil from a lease tank to pumping station without the use of mechanical means; a line that transports liquid from one elevation to a lower elevation by the force of gravity alone.

GRAVITY MAPS
Results of reconnaissance gravity surveys; display of gravity measurements taken in an area. *See* Gravimeter.

GRAVITY SEGREGATION
The separation of water from oil, or heavy from lighter hydrocarbons by the

force of gravity, either in the producing zone or by gravity in the separators after production; the stratification of gas, oil, and water according to their densities.

GRAVITY STRUCTURE
An offshore drilling and production platform made of concrete and of such tremendous weight that it is held securely on the ocean bottom without the need for piling or anchors. One of the world's largest gravity structures is located off the Scottish coast in the North Sea. Its general configuration is that of a column mounted on a large circular base which has storage for 1 million barrels of crude. The base is 450 feet in diameter; the column is 180 feet in diameter and the overall height of the structure is about 550 feet.

GREASE
(1) A lubricating substance (solid or semi-solid) made from lubricating oil and a thickening agent. The lube oils may be very light or heavy cylinder oils; the thickening agent (usually soaps) may be any material that when mixed with oil will produce a grease structure. (2) Colloquial for crude oil.

GRIEF STEM
Kelly joint; the top joint of the rotary drill string that works through the square hole in the rotary table. As the rotary table is turned by the drilling engines, the grief stem and the drill pipe are rotated. Grief stems are heavy, thick-walled tubular pieces with squared shoulders that are made to fit into the hole in the rotary table.

GRIND OUT
Colloquial for centrifuge; to test samples of crude oil or other liquid for suspended material—water, emulsion, sand—by use of a centrifuge machine.

GROSS PRODUCTION TAX
A severance tax (q.v.); a tax usually imposed by a state, at a certain sum per unit of mineral removed (barrels of oil, thousands or millions of cubic feet of gas).

GROUND-SEAT UNION
A pipe coupling made in two parts; one half is convex, the other half concave in shape, and both ground to fit. A threaded ring holds the halves together, pressure tight. Used on small-diameter piping.

GROUT
(1) A concrete mixture used to fill in around piling, caissons, heavy machinery beds, and foundation work. (2) To stabilize and make permanent. Grout is usually a thin mixture that can be worked into crevices and beneath and around structural forms.

GROWLER BOARD
See Lazy board

GRUB STAKE AGREEMENT
An agreement whereby one person undertakes to prospect for oil and agrees to hand over to the person who furnishes money or supplies a certain proportionate interest in the oil discovered. This type of agreement is common for solid minerals but is not often used in oil prospecting.

G.S.A.
Geological Society of America

GUMBO
A heavy, sticky mud formed downhole by certain shales when they become wet from the drilling fluid.

GUN PERFORATION
A method of putting holes in a well's casing downhole in which explosive charges lowered into the hole propel steel projectiles through the casing wall. (Casing is perforated to permit the oil from the formation to enter the well.)

GUNK
The collection of dirt, paraffin, oil, mill scale, rust, and other debris that is cleaned out of a pipeline when a scraper or a pig is put through the line.

GUSHER
A well that comes in with such great pressure that the oil blows out of the wellhead and up into the derrick, like a geyser. With improved drilling technology, especially the use of drilling mud to control downhole pressures, gushers are rare today. *See* Blowout.

GUY WIRE
A cable or heavy wire used to hold a pole or mast upright. The end of the guy wire is attached to a stake or a deadman (q.v.).

H

HALF-SOLE
A metal patch for a corroded section of pipeline. The patch is cut from a length of pipe of the same diameter as the one to be repaired. Half-soles can be from six to 12 feet in length, and are placed over the pitted or

corroded section of the pipe and welded in place with a bead around the entire perimeter of the half-sole.

HAND MONEY
See Earnest money

HANDY
Hand-tight; a pipe connection or nut that can be unscrewed by hand.

HANG A WELL OFF
To disconnect the pull-rod line from a pumping jack or pumping unit being operated from a central power (q.v.).

HANG THE RODS
To pull the pump rods out of the well and hang them in the derrick on rod hangers. On portable pulling units, the rods are hung outside the derrick.

HARDWARE
(1) Electronic and mechanical components of a computer system, e.g. storage drums, scanners, printers, computers. (2) Mechanical equipment, parts, tools.

HARDWARE CLOTH
A type of galvanized metal screen that can be bought in hole sizes, e.g. ⅛, ¼, ½-inch, etc. The holes are square.

HATCH
An opening in the top of a tank or other vessel through which inspections are made or samples taken; a gauge hatch.

HATCHET KNOT
A knot that defies untying and so must be cut; a granny knot.

HAT-TYPE FOUNDATION
A metal base or foundation in the shape of an inverted, rectangular cake pan. Hat-type foundations are used for small pumps and engines or other installations not requiring solid, permanent foundations.

HAUL ASS
An inelegant term meaning to leave a place with all haste; vamos; split.

HAY RAKE
Another name for the finger board (q.v.) in the derrick of an oil well.

HEAD
See Hydrostatic head

HEAD WELL
A well that makes its best production by being pumped or flowed intermittently.

HEADACHE!
A warning cry given by a fellow worker when anything is accidentally dropped or falls from overhead toward another worker.

HEADACHE POST
A frame over a truck cab that prevents pipe or other material being hauled from falling on the cab; a timber set under the walking beam to prevent it from falling on the drilling crew when it is disconnected from the crank and pitman (q.v.).

HEADER
A large-diameter pipe into which a number of smaller pipes are perpendicularly welded or screwed; a collection point for oil or gas gathering lines. *See* Manifold.

HEADING
An intermittent or unsteady flow of oil from a well. This type of flow is often caused by a lack of gas to produce a steady flow thus allowing the well's tubing to load up with oil until enough gas accumulates to force the oil out.

HEATER
(1) An installation used to heat the stream from high-pressure gas and condensate wells (especially in winter) to prevent the formation of hydrates, a residue which interferes with the operation of the separator. (2) A refinery furnace.

HEATER, PIPELINE
An installation fitted with heating coils or tubes for heating certain crude oil to keep it "thin" enough to be pumped through a pipeline. Crudes with high pour points (congealing at ordinary temperatures) must be heated before they can be moved by pipeline.

HEAT TAPE
An electrical heating element made in the form of an insulated wire or tape used as a tracer-line to provide heat to a pipeline or instrument piping. The heat tape is held in direct contact with the piping by a covering of insulation.

HEAVE COMPENSATOR
A type of snubber-shock absorber on a floating, drilling platform or drill ship that maintains the desired weight on the drill bit as the unstable platform heaves on ocean swells. Some compensators are made with massive counterweights others have hydraulic systems to keep the proper weight on the bit constant. Without compensators, the bit would be lifted off bottom as the platform rose on each swell.

HEAVY BOTTOMS
A thick, black residue left over from the refining process after all lighter

fractions are distilled off. Heavy bottoms are used for residual fuel and for asphalt.

HEAVY ENDS
In refinery parlance, heavy ends are the heavier fractions of refined oil—fuel oil, lubes, paraffin, and asphalt—remaining after the lighter fractions have been distilled off. *See* Light ends.

HEAVY FUEL OIL
A residue of crude oil refining processes. The product remaining after the lighter fraction—gasoline, kerosine, lubricating oils, wax, and distillate fuels—have been extracted from the crude; residual fuel oil.

HEAVY-METAL
Spent-uranium or tungsten. Heavy-metal is used to make drill tools to add weight to the drill assembly. Drill collars made of heavy-metal weigh twice as much as those made of steel, and are used to stabilize the bit and to force it to make a straighter hole, with less deviation from the vertical.

HIDE THE THREADS, TO
To make up (tighten) a joint of screw pipe until all the threads on the end of the joint are screwed into the collar, hiding the threads and making a leak-proof connection.

HIGH
A geological term for the uppermost part of an inclined structure where the likelihood of finding oil is considered to be the greatest.

HIGH BOTTOM
A condition in a field stock tank when BS&W (basic sediment and water) has accumulated at the bottom of the tank to a depth making it impossible to draw out the crude oil without taking some of the BS with it into the pipeline. When this condition occurs the operator (lease pumper) must have the tank cleaned before the pipeline company will run the tank of oil.

HI-BOY
A skid-mounted or wheeled tank with a hand-operated pump mounted on top used to dispense kerosine, gasoline, or lubricating oil to small shops and garages.

HIGH-PRESSURE GAS INJECTION
Introduction of gas into a reservoir in quantities exceeding the volumes produced in order to maintain reservoir pressure high enough to achieve mixing between the gas and reservoir oil. *See* Solution gas.

HISTORY OF A WELL
A written account of all aspects of the drilling, completion, and operation of a well. (Well histories are required in some states.)

HITTING THE HOOKS
Working on a pipeline, screwing in joints of pipe using pipe tongs; an expression used by the tong crew of a pipeline gang. The tong crews on large diameter screw-pipelines (up to about 12-inch pipe) hit the hooks in perfect rhythms. With three sets of tongs on the joint being screwed in, each large tong, run by two or three men, made a stroke every third beat of the collar pecker's hammer (q.v.) until the joint was nearly screwed in. Then the three tongs, with all six or nine men, hit together to "hide the threads," to tighten the joint the final and most difficult round.

HHP
Hydraulic horsepower; a designation for a type of very-high-pressure plunger pump used in downhole operations such as cementing, hydro-fracturing, and acidizing.

HOISTING DRUM
A powered reel holding rope or cable for hoisting and pulling; a winch. *See* Draw works.

HOLD DOWN/HOLD UP
Oscillating anchoring devices or supports for a shackle-rod line to hold the rod line to the contour of the land it traverses. The devices are timbers or lengths of pipe hinged to a deadman or overhead support at one end, the other end attached to and supporting the moving rod line.

HOLE OPENER
A type of reamer used to increase the diameter of the well bore below the casing. The special tool is equipped with cutter arms that are expanded against the wall of the hole and by rotary action reams a larger diameter hole.

HOLIDAYS
Breaks or flaws in the protective coating of a joint of line pipe. Holidays are detected by electronic testing devices as the pipe is being laid. When detected, the breaks are manually coated. *See* Jeeping.

HOOK BLOCK
One or more sheaves or pulleys in a steel frame with a hook attached. *See* Traveling block.

HOOK-LOAD CAPACITY
The maximum weight or pull a derrick and its lines, blocks, and hook are designed to handle. A rating specification for a drilling rig. (Some large rigs have a hook-load capacity of two million pounds.)

HOOKS
Pipe-laying tongs named for the shape of the pipe-gripping head of the scissor-like wrench.

HOOK, THE
The hook attached to the frame of the rig's traveling block (q.v.) and which engages the bail of the swivel in drilling operations. *See* Hook-load capacity.

HOOK UP
To make a pipeline connection to a tank, pump, or a well. The arrangement of pipes, nipples, flanges, and valves in such a connection.

HORIZON
A zone of a particular formation; that part of a formation of sufficient porosity and permeability to form a petroleum reservoir. *See* Pay zone.

HORIZONTAL INTEGRATION
Refers to the condition in which a diversified company has resources or investments other than its principle business, and from which it makes a profit. Specifically, an oil company is said to be horizontally integrated when, besides oil and gas holdings, it owns coal deposits, is into nuclear energy, oil shale or geothermal energy. *See* Vertical integration.

HORSE FEED
An old oil field term for unexplainable expense account items in the days of the teamster and line ride who were given an allowance for horse feed. Expenses that needed to be masked in anonymity were simply listed "horse feed."

HORSEHEAD
The curved guide or head piece on the well-end of a pumping jack's walking beam. The metal guide holds the short loop of cable (the bridle) attached to the well's pump rods.

HORSEPOWER
A unit of power equivalent to 33,000 foot-pounds a minute or 745.7 watts of electricity.

HORTONSPHERE
A spherical tank for the storage, under pressure, of volatile petroleum products, e.g. gasoline and LP-gases; also Hortonspheroid, a flattened spherical tank, somewhat resembling a tangerine in shape.

HOT-FLUID INJECTION
A method of thermal oil recovery, in which hot fluid (water, gas, or steam) is injected into a formation to increase the flow of low-gravity crude to production wells.

HOT FOOTING
Installing a heater at the bottom of an input well to increase the flow of heavy crude oil from the production wells. *See* Hot-fluid injection.

HOT-HEAD ENGINE
A hot-plug engine (q.v.); a "semi-diesel."

HOT OIL
Oil produced in violation of state regulations or transported interstate in violation of Federal regulations.

HOT PASS
A term describing a "bead" or course of molten metal laid down in welding a pipeline. The hot pass is the course laid down on top of the stringer bead, which is the first course in welding a pipeline. *See* Pipeline welding.

HOT-PLUG ENGINE
A stationary diesel-cycle engine that is started by first heating an alloy-metal plug in the cylinder head that protrudes into the firing chamber. The hot plug assists in the initial ignition of the diesel fuel until the engine reaches operating speed and temperature. Afterwards the plug remains hot, helping to provide heat for ignition; hot-tube engine; hot head. *See* Semi-diesel.

HOT TAPPING
Making repairs or modifications on a tank, pipeline, or other installation without shutting down operations. *See* Tapping and plugging machine.

HOUSE-BRAND (GASOLINE)
An oil company's regular gasoline; a gasoline bearing the company's name.

H_2S
Hydrogen sulfide (q.v.).

HUMPHREYS, DR. R. E.
A petroleum chemist who worked with Dr. W. M. Burton in developing the first commercially successful petroleum cracking process using heat and pressure.

HUNDRED-YEAR STORM CONDITIONS
A specification for certain types of offshore installations—production and drilling platforms, moorings, and offshore storage facilities—is that they be built to withstand winds of 125 miles an hour and "hundred-year storm conditions;" the biggest blow on record.

HURRY-UP STICK
The name given to the length of board with a hole in one end which the cable-tool driller used to turn the T-screw at the end of the temper screw (q.v.) when the walking beam was in motion. This enabled the driller to perform the job of letting out the drilling line easily and rapidly.

HYDRAULIC FRACTURING
A method of stimulating production from a formation of low permeability by inducing fractures and fissures in the formation by applying very high fluid-pressure to the face of the formation, forcing the strata apart. Various patented techniques, using the same principle, are employed by oil field service companies.

HYDROCARBONS
Organic chemical compounds of hydrogen and carbon atoms. There are a vast number of these compounds and they form the basis of all petroleum products. They may exist as gases, liquids, or solids. An example of each is methane, hexane, and asphalt.

HYDROCRACKING
A refining process for converting middle-boiling or residual material to high-octane gasoline, reformer charge stock, jet fuel and/or high-grade fuel oil. Hydrocracking is an efficient, relatively low-temperature process using hydrogen and a catalyst. The process is considered by some refiners as a supplement to the basic catalytic cracking process.

HYDRODYNAMICS
A branch of science that deals with the cause and effect of regional subsurface migration of fluids.

HYDROGEN SULFIDE (H_2S)
An odorous and noxious compound of sulfur found in "sour" gas. *See* Sour gas.

HYDROMETER
An instrument designed to measure the specific gravity of liquids; a glass tube with a weighted lower tip that causes the tube to float partially submerged. The API gravity of a liquid is read on a graduated stem at the point intersected by the liquid.

HYDROSTATIC HEAD
The height of a column of liquid; the difference in height between two points in a body of liquid.

HYDROSTATIC TESTING
Filling a pipeline or tank with water under pressure to test for tensile strength, its ability to hold a certain pressure without rupturing.

HYPERBARIC WELDING (EXCESSIVE-PRESSURE WELDING)
Welding on the sea bottom "in the dry" but under many atmospheres of pressure (compression). In hyperbaric welding of under-sea pipelines, a large frame is lowered into the water and clamped to the pipeline. Then an open-bottomed, box-like enclosure is placed in the center of the frame

over the pipe. Power lines and life-support umbilicals are connected to the box. The sea water is displaced with breathing-gas mixtures for the diver-welders permitting them to do their work in the dry but high-pressure atmosphere.

I

IADC
International Association of Drilling Contractors

I.C.C.
Interstate Commerce Commission

ID
Inside diameter—of a pipe or tube; initials used in specifying pipe sizes, e.g. 3½-inch ID; also OD, outside diameter, e.g. 5-inch OD.

IDIOT STICK
A shovel, or other digging tool not requiring a great deal of training to operate.

IDLER GEAR OR WHEEL
A gear so called because it is usually located between a driving gear and a driven gear, transmitting the power from one to the other. It also transmits the direction of rotation of the driving to the driven gear. Without the idler or the intermediate gear, the driving gear by directly meshing with the driven gear reverses the direction of rotation. Idler wheels or pulleys are also used for tightening belts or chains or to maintain a uniform tension on them.

IGNITION MAGNETO
An electric current generator used on stationary engines in the field. A magneto is geared to the engine and once the engine is started either by hand cranking or by a battery starter, the magneto continues to supply electric current for the ignition system. Current is produced by an armature rotating in a magnetic field created by permanent magnets.

IGNORANT END
The heaviest end of a tool or piece of equipment to be carried or operated.

I.H.P.
Indicated horsepower

IMPACT WRENCH
An air-operated wrench for use on nuts and bolts of large engines, valves, and pumps. Impact wrenches have taken the place of heavy end-wrenches and sledge hammers in tightening and loosening large nuts. A small version of the impact wrench is the air-operated automobile lug-wrench used at modern service stations and garages.

IN-LINE EQUIPMENT
Pumps, separators, heat exchanges integral to a process or processing chain; in the line, not auxiliary or only supporting.

INCLINOMETER
An instrument used down hole to determine the degree of deviation from the vertical of a well bore at different depths.

INDEPENDENT PRODUCER
(1) A person or corporation that produces oil for the market, having no pipeline system or refinery. (2) An oil-country entrepreneur who secures financial backing and drills his own well.

INDICATED HORSEPOWER (I.H.P.)
Calculated horsepower; the power developed within the cylinder of an engine which is greater than the power delivered at the drive shaft by the amount of mechanical friction that must be overcome. *See* Brake horse-power.

INDICATOR PASTE, GASOLINE
A viscous material applied to a steel gauge line or gauge pole that changes color when it comes in contact with gasoline, making it easy for the gauger to read the height of gasoline in the tank.

INDICATOR PASTE, WATER
A paste material applied to a steel gauge line or wooden gauge pole that changes color when immersed in water. It is used to detect the presence of water in a tank of oil.

INDUSTRIAL GAS
Gas purchased for resale to industrial users.

INFILL DRILLING
Wells drilled to fill in between established producing wells on a lease; a drilling program to reduce the spacing between wells in order to increase production from the lease.

INFLUENT
The flow of liquids or gas into a vessel or equipment. *See* Effluent.

INGAA
Interstate Natural Gas Association of America

INNAGE GAUGE
A measure of the quantity of oil in a tank calculated on the basis of the depth of oil in the tank; the most common method of gauging a tank. *See* Outage gauge.

INNOVATOR'S ROYALTY
A type of overriding royalty paid to the person instrumental in bringing a company to a concession from a foreign government; British: a fixer's royalty. *See* Overriding royalty.

INPUT WELL
A well used for injecting water or gas into a formation in a secondary recovery or pressure maintenance operation.

IN-SITU COMBUSTION
A technique used in some locations for recovering oil of low gravity and high viscosity from a field when other primary methods have failed. Essentially, the method involves the heating of the oil in the formation by igniting the oil (burning it in place) and keeping the combustion alive by pumping air down hole. As the front of burning oil advances, the heat breaks down the oil into coke and light oil. And as the coke burns, the lighter, less viscous oil is pushed ahead to the well bores of the producing wells.

INSPECTION PLATE
A flat metal plate fitted with a gasket and bolted over an opening in the gearbox of a pump or the crankcase of an engine. By removing the plate an inspection of the gears or crank and connecting-rod bearings can be made. On large, multi-cylinder engines, inspection windows are large enough to permit a mechanic to enter the crankcase to inspect or "change out" a bearing.

INTANGIBLE DRILLING COSTS
Expenditures incurred by an operator for labor, fuel, repairs, hauling, and supplies used in drilling and completing a well for production.

INTANGIBLES
Short for intangible drilling costs (q.v.).

INTEGRATED OIL COMPANY
A company engaged in all phases of the oil business, i.e., production, transportation, refining, and marketing; a company that handles its own oil from wellhead to gasoline pump.

INTERFACE
The point or area where two dissimilar products or grades of crude oil meet in a pipeline as they are pumped, one behind another.

INTERMEDIATE STRING
See Casing

INTERRUPTABLE GAS
A gas supply, usually to industrial plants and large commercial firms, that can be curtailed or interrupted during emergencies or supply shortages in order to maintain service to domestic customers.

INTERSTATE OIL COMPACT
A compact between oil-producing states negotiated and approved by Congress in 1935, the purpose of which is the conservation of oil and gas by the prevention of waste. The Compact provides no power to coerce but relies on voluntary agreement to accomplish its objectives. Originally, there were six states as members; today, there are nearly 30.

IOCC
Interstate Oil Compact Commission

IOSA
International Oil Scouts Association

IPAA
Independent Petroleum Association of America

ISO-
A prefix denoting similarity. Many organic substances, although composed of the same number of the same atoms, appear in two, three, or more varieties or isomers which differ widely in physical and chemical properties. In petroleum fractions there are many substances that are similar, differing only in specific gravity, for example, iso-octane, iso-butane, iso-pentane, and many other isomers.

ISOMERIZATION
A refinery process for converting chemical compounds into their isomers, i.e., rearranging the structure of the molecules without changing their size or chemical composition.

ISOMERS
Compounds having the same composition and the same molecular weight but differing in properties.

ISOPACHOUS MAP
A geological map; a map that shows the thickness and conformation of underground formations; used in determining underground oil and gas reserves.

ISOPENTANE
A high-octane blending stock for automotive gasoline.

ISOTHERMAL
At constant temperature. When a gas is expanded or compressed at a constant temperature, the expansion or compression is isothermal. Heat must be added to expanding gas and removed from compressing gas to keep it isothermal.

J

J-4 FUEL
A designation for highly-refined kerosine used as fuel for jet engines.

JACK
An oil well pumping unit powered by a gasoline engine, electric motor, or rod line from a central power. The pumping jack's walking beam provides the up and down motion to the well's pump rods.

JACK BOARD
A wood or metal prop used to support a joint of line pipe while another joint is being screwed into it. Jack boards have metal spikes inserted at intervals to support the pipe at different levels.

JACK RABBIT
A device that is put through casing or tubing before it is run to make certain it is the proper size inside and outside; a drift mandrel.

JACK-KNIFE RIG
A mast-type derrick whose supporting legs are hinged at the base. When the rig is to be moved, it is lowered or laid down intact and transported by truck.

JACK-UP RIG
A barge-like, floating platform with legs at each corner that can be lowered to the sea bottom to raise or "jack up" the platform above the water. Towed to location offshore, the legs of the jack-up rig are in a raised position, sticking up high above the platform. When on location, the legs are run down hydraulically or by individual electric motors.

JAM NUT
A nut used to jam and lock another nut securely in place; the second and locking nut on a stud bolt. After the first nut is threaded and tightened on a stud, a second nut is tightened down on the first nut to prevent it from working loose.

JARS
A tool for producing a jarring impact in cable-tool drilling, especially when the bit becomes stuck in the hole. Cable-tool jars (part of the drill string) are essentially a pair of elongated, interlocking steel links with a couple of feet of "play" between the links. When the drilling line is slacked off, the upper link of the jars move down into the lower link. When the line is suddenly tightened the upper link moves upward engaging the lower link with great force that usually frees the stuck bit. See Bumper sub, fishing.

JEEPING
Refers to the operation of inspecting pipe coating with the aid of electronic equipment. An indicator-ring is passed over the pipe which carries an electric charge. If there is a break or holiday (q.v.) in the protective coating, a signal is transmitted through the indicator-ring to an alarm.

JERKER
A small, one-cylinder plunger pump. See Pump, jerker.

JERK LINE
A line which connects the bandwheel crank to the drilling cable. As the crank revolves, the drilling line is jerked (pulled up and released suddenly) providing an up and down motion to the spudding tools on a cable tool rig.

JET FUEL
A specially refined grade of kerosine used in jet propulsion engines.

JETTED-PARTICLE DRILLING
A method of drilling in hard rock formations using steel pellets forced at high velocity from openings in the bottom of the drill bit.

JETTING
Injecting gas into a subsurface formation for the purpose of maintaining reservoir pressure.

JOCKEY
An experienced and proficient driver of large trucks or earth-moving equipment.

JOINT
A length of pipe, casing, or tubing usually from 20 to 30 feet long. On drilling rigs, drill pipe and tubing are run the first time (lowered into the hole) a joint at a time; when pulled out of the hole and stacked in the rig, they are usually pulled two, three, or four at a time depending upon the height of the derrick. These multiple-joint sections are called stands (q.v.).

JOINT ADVENTURE
See Joint venture

JOINT VENTURE

A business or enterprise entered into by two or more partners. Joint venture leasing is a common practice. Usually the partner with the largest interest in the venture will be the operator. *See* Consortium.

JOURNAL

That part of a rotating shaft that rests and turns in the bearing; the weight-bearing segment of the shaft.

JOURNAL BOX

A metal housing that supports and protects a journal bearing. *See* Journal.

JUMBO BURNER

A flare used for burning "waste" gas produced with oil when there is no ready market or the supply is too small or temporary to warrant a pipeline.

JUMBOIZING

A technique used to enlarge an oil tanker's carrying capacity by cutting the vessel in two amidships and inserting a section between the halves.

JUNK BASKET

A type of fishing tool used to retrieve objects lost in the bore hole or down the casing.

K

KELLY COCK

A blow-out preventer built inside a three-foot section of steel tubing inserted in the drill string above the kelly. A kelly cock is also inserted in the string below the kelly joint in some instances.

KELLY JOINT

The first and the sturdiest joint of the drill column; the thick-walled, hollow steel forging, with two flat sides and two rounded sides, that fit into a square hole in the rotary table which rotates the kelly joint and the drill column. Attached to the top of the kelly or grief stem (q.v.) is the swivel and mud hose.

KELLY SAFETY VALVE

See Kelly cock

KEROGEN
A bituminous material occurring in certain shales which yield a type of oil when heated. *See* Kerogen shales.

KEROGEN SHALES
Commonly called oil shales, kerogen shales contain material neither petroleum nor coal but an intermediate bitumen material with some of the properties of both. Small amounts of petroleum are usually associated with kerogen shales but the bulk of the oil is derived from heating the shale to about 660°F. Kerogen is identified as a pyrobitumen.

KEROSINE, RAW
Kerosine-cut from the distillation of crude oil, not treated or "doctor tested" to improve odor and color.

KEY
(1) A tool used in pulling or running sucker rods of a pumping oil well; a hook-shaped wrench that fits the square shoulder of the rod connection. Rod wrenches are used in pairs; one to hold back-up and the other to break out and unscrew the rod. (2) A slender metal piece used to fasten a pulley wheel or gear onto a shaft. The key fits into slots (keyways) cut in both the hub of the wheel and the shaft.

KEYWAY
A groove or slot in a shaft or wheel to hold a key (q.v.).

KEYSEAT
A section of the well bore deviating abruptly from the vertical causing drilling tools to hang up; a shoulder in the bore hole.

KEYSEATING
A condition downhole when the drill collar or another part of the drill string becomes wedged in a section of crooked hole, particularly a dog leg which is an abrupt deviation from the vertical or the general direction of the hole being drilled.

KICK
Pressure from down hole in excess of that exerted by the weight of the drilling mud, causing loss of circulation. If the gas pressure is not controlled by increasing the mud weight, a kick can violently expel the column of drilling mud resulting in a blowout.

KICKING DOWN A WELL
A primitive method of drilling a shallow well using man power (leg power). In oil's very early days, a pole made from a small tree was used to support the drilling line and bit in the hole. The driller with his foot in a stirrup attached to the line would kick downward causing the pole to bend and

the bit to hit the bottom of the hole. The green sapling would spring back, lifting the bit ready for another "kick" by the driller.

KILL AND CHOKE LINES
Lines connected to the blowout preventer stack through which drilling mud is circulated when the well has been shut in because excessive pressure downhole has threatened a blowout. Mud is pumped through the kill line and is returned through the choke line, bypassing the closed valves on the BOP. When the mud has been heavied up to overcome down hole pressure, drilling can proceed.

KILL A WELL
To overcome downhole pressure in a drilling well by the use of drilling mud or water. One important function of drilling mud is to maintain control over any downhole gas pressures that may be encountered. If gas pressure threatens to cause loss of circulation or a blowout, drilling mud is made heavier (heavied-up) by the addition of special clays or other material. *See* Kick.

KILLER WELL
A directional well drilled near an out-of-control well to "kill" it by flooding the formation with water or mud. Wells that have blown out and caught fire are often brought under control in this manner if other means fail.

KNOCK-OFF POST
A post through which a rod line moves as it operates a pumping jack. When the well is to be hung off (shut down), a block is inserted between the rod-line hook and the knock-off post which interrupts the line's forward movement putting slack in the line so that the hook may be disengaged.

KNOCKOUT
A tank or separator vessel used to separate or "knock out" water from a stream of oil.

KNUCKLE-BUSTER
A wrench so worn or of such poor quality that it will not hold when under the strain of heavy work.

KNUCKLE-JOINT
A universal joint (q.v.); a type of early drilling tool hinged on a movable joint so that the drill could be deflected at an angle from the vertical.

L

LACT
Lease Automatic Custody Transfer. *See* Automatic custody transfer.

LANDED COST (OF OIL)
The cost of a barrel of imported oil offloaded at a U.S. port. Landed cost includes all foreign taxes and royalties plus cost of transportation.

LANDING CASING
Lowering a string of casing into the hole and setting it on a shoulder of rock at a point where the diameter of the bore hole has been reduced. The beginning of the smaller diameter hole forms the shoulder on which the casing is landed.

LANDMAN
A person whose primary duties are managing an oil company's relations with its landowners. Such duties include securing of oil and gas leases, lease amendments, and other agreements.

LANDOWNER ROYALTY
A share of the gross production of the oil and gas on a property by the landowner without bearing any of the cost of producing the oil or gas. The usual landowner's royalty is one-eighth of gross production.

LAP-WELDED PIPE
Line pipe or casing made from a sheet of steel which is formed on a mandrel. The two edges, tapered to half normal thickness, are lapped over and welded. *See* Seamless pipe.

LATERAL LINES
Pipelines that tie into a trunk line; laterals are of smaller diameter and are laid as part of a gathering system or a distribution system. In an oil field, laterals bring oil or gas from individual leases or tank batteries to the booster station and the trunk line.

LAY DOWN THE TUBING
To pull the tubing from the well, a joint at a time, and remove it from the derrick floor to a nearby horizontal pipe rack. As each joint is unscrewed from the string, the lower end of the joint is placed on a low cart and pulled out to the rack as the driller lowers the pipe which is held up by the elevators.

LAY TONGS
See Pipe tongs

LAY-BARGE
A shallow-draft, barge-like vessel used in the construction and laying of under-water pipelines. Joints of line pipe are welded together and paid out over the stern of the barge as it is moved ahead. Lay-barges are used in swampy areas, in making river crossings, and laying lines to offshore installations.

LAY-DOWN RACK
A storage area for tubing and drill pipe that are removed from a well and laid down rather than set back and racked vertically in the derrick.

LAZY BENCH
A bench on which workers, when not working, may rest. A perch from which a work operation may be observed by workers or guests.

LAZY BOARD
A stout board with a handle used to support the end of a pipeline while another length of pipe is screwed into it. On small lines, the man operating the lazy board or "granny" board usually handles the back-up wrench which holds one joint of pipe firm while another joint is being screwed in.

LB/LB
Pound per pound. In a refining process, the ratio of ingredients to be mixed or introduced to the process.

LBS-H₂O MM
Pounds of water per million standard cubic feet (MMSCF) of natural gas. The designation of water content for large volumes of gas. See PPM/WT

LCCV
Large crude carrying vessel; tankers from 100,000 to 500,000 dead-weight tons capable of transporting 2.5 to 3.5 million barrels of oil in one trip. Cruising speed of LCCVs is 12 to 18 knots; overall length, about 1,200 feet; draft when fully loaded: more than 80 feet.

LEAD LINES
Lines through which production from individual wells is run to a lease tank battery.

LEAN GAS
Natural gas containing little or no liquefiable hydrocarbons. See Wet gas.

LEAN OIL
The absorbent oil in a gasoline absorption plant from which the absorbed gasoline fractions have been removed by distillation. Before distillation to remove gasoline fractions, the oil is referred to as "fat oil" (q.v.).

LEASE
(1) The legal instrument by which a leasehold is created in minerals. A

contract that, for a stipulated sum, conveys to an operator the right to drill for oil or gas. The oil lease is not to be confused with the usual lease of land or a building. The interests created by an oil-country lease are quite different from a reality lease. (2) The location of production activity; oil installations and facilities; location of oil field office, tool house, garages.

LEASE BROKER
A person whose business is securing leases for later sale in the hope of profit. Lease brokers operate in areas where survey or exploration work is being done.

LEASE CONDENSATE
Liquid hydrocarbons produced with natural gas and separated from the gas at the well or on the lease. *See* Condensate.

LEASE HOUND
Colloquial term for a person whose job is securing oil and gas leases from landowners for himself or a company for which he works. *See* Land man.

LEFT-HAND THREAD
A pipe or bolt thread cut to be turned counter-clockwise in tightening. Most threads are right-hand, cut to be tightened by turning clockwise. Nipples with one kind of thread on one end, another on the other end are referred to as "bastard (q.v.) nipples."

LEGAL SUBDIVISION
Forty acres; one-sixteenth of a section (square mile).

LEGS, OIL DERRICK
The four corner-members of the rig, held together by sway braces and girts.

LENS
A sedimentary deposit of irregular shape surrounded by impervious rock. A lens of porous and permeable sedimentary rock may be an oil-producing area.

LESSEE
The person or company entitled, under a lease, to drill and operate an oil or gas well.

LIFTING
(1) Refers to tankers and barges taking on cargoes of oil or refined product at a terminal or transshipment point. (2) Producing an oil well by mechanical means: pump, compressed air, or gas.

LIFTING COSTS
The costs of producing oil from a well or a lease.

LIGHT ENDS
The more volatile products of petroleum refining; e.g. butane, propane, gasoline.

LIGHT PLANT
An early-day term for an installation on a lease or at a company camp (q.v.) that provided electricity for lighting and small appliances. The light plant often was simply a belt-driven D.C. generator run off one of the engines at a pipeline pumping station or a pumping well's engine. The lights "surged" with the power strokes of the engines, and went out when an engine "went down," but the lights were far better than gas lights—or none at all.

LIME
Colloquial for limestone.

LINER
In drilling, a length of casing used downhole to shut off a water or gas formation so drilling can proceed. Liners are also used to case a "thief zone" (q.v.) where drilling fluid is being lost in a porous formation. A liner is also a removable cylinder used in reciprocating pumps and certain types of internal-combustion engines; a sleeve.

LINE-UP CLAMPS
A device that holds the ends of two joints of pipe together and in perfect alignment for welding. Line-up clamps operate on the outside of the pipe and are used on smaller diameter line pipe. Large-diameter pipe—20 to 36-inch and over—are aligned by internal, hydraulically operated mandrel-like devices.

LINE, GAS SALES
Merchantable natural-gas line from a lease or offshore production-processing platform carrying gas that has had water and other impurities removed; a line carrying pipeline gas (q.v.).

LINE, OIL SALES
Merchantable crude-oil line from a lease or offshore, production-processing platform carrying oil that has had water and other impurities removed; a line transporting pipeline oil, (q.v.).

LINKAGE
A term used to describe an arrangement of interconnecting parts—rods, levers, springs, joints, couplings, pins—that transmit motion, power, or exert control.

LIQUEFIED PETROLEUM GAS
Butane, propane, and other light ends (q.v.) separated from natural

gasoline or crude oil by fractionation or other processes. At atmospheric pressure, liquefied petroleum gases revert to the gaseous state.

LIQUID HYDROCARBONS
Petroleum components that are liquid at normal temperatures and atmospheric pressure.

LITER
A metric unit of volume; 1.057 U.S. quarts; 61.02 cubic inches.

LITHOLOGY
(1) The study and identification of rocks. (2) The character of a rock formation.

LIVE OIL
Crude oil which contains dissolved natural gas when produced.

LNG
Liquefied natural gas; natural gas converted to a liquid state by pressure and severe cooling.

LNGC
Liquefied natural gas carrier; a specially designed ocean-going vessel for transporting liquefied natural gas.

LOADING ARMS
Vertical standpipes with swivel-jointed arms that extend to a tanker or barge's deck connections for loading crude oil or products.

LOADING RACK
An elevated walkway that supports vertical filling lines and valves for filling tank cars from the top.

LOAD OIL
Oil of any kind put back into a well for any purpose; e.g. hydraulic fracturing, shooting, or swabbing.

LOCATION
The well site; the place where a well is to be drilled or has been drilled; a well spacing unit, e.g. "Two locations south of the discovery well. . . . "

LONG STRING
See Production string

LOOPING A LINE
The construction of a pipeline parallel to an existing line, usually in the same right-of-way, to increase the throughput capacity of the system; doubling a pipeline over part of its length, with the new section tied into the original line.

LOSE RETURNS, TO
Refers to a condition in which less drilling mud is being returned from downhole than is being pumped in at the top. This indicates that mud is being lost in porous formations, crevices, or a cavern.

LOSS OF CIRCULATION
A condition that exists when drilling mud pumped into the well through the drill pipe does not return to the surface. This serious condition results from the mud being lost in porous formations, a crevice or a cavern penetrated by the drill.

LOWBOY
A low-profile, flat-bed trailer with multiple axles (6 to 10) for transporting extra-heavy loads over relatively short distances. The many wheels and axles serve to spread the weight of the trailer and its load over a large area to avoid damaging streets and highways. The low bed makes it easier to load and unload the heavy equipment it was designed to move.

LOWER IN
To put a completed pipeline in the ditch. This is done with side-boom tractors that lift the pipe in slings and carefully lower it into the ditch.

LPG
Liquefied Petroleum Gas

LUBE OIL
Short for lubricating oil or lubricant. Also lube and lubes.

LUBRICATING OIL, MULTI-GRADE
Specially formulated lubricating oil that flows freely in cold weather, and in the heat of engine operation, maintains sufficient viscosity or "body" to properly lubricate the engine; e.g. 10-30 SAE.

LUBRICATOR, MUD
A temporary hook-up of pipes and valves for introducing additional, heavy drilling mud into the well bore to control gas pressure. Through one or two joints of large-diameter casing attached atop the wellhead, the heavy mud is fed into the well bore, against pressure, as through a lubricator.

LUBRICATOR, OIL
A small, box-like reservoir containing a number of gear-operated pumps. The individual pumps, working in oil, measure out a few drops at a time into small, copper lines that distribute the lubricant to the bearings.

LUCAS, CAPT. ANTHONY F.
It was Capt. Lucas' Spindletop gusher in 1901 (75,000 bbls/day) that ushered in the modern oil age of large oil companies. John H. Galey and

James M. Guffey owned the Spindletop gusher located near Beaumont, Texas.

LXT UNIT
A low-temperature separator; a mechanical separator which uses refrigeration obtained by expansion of gas from high pressure to low pressure to increase recovery of gas-entrained liquids.

M

MAGNETO
See Ignition magneto

MAGNETOMETER
An instrument for measuring the relative intensity of the earth's magnetic effect. Used to detect rock formations below the surface; an instrument used by geophysicists in oil exploration work.

MAIN LINE
Trunk line; a large-diameter pipeline into which smaller lines connect; a line that runs from an oil-producing area to a refinery.

MAKE A HAND
To be a good worker.

MAKE IT UP
To screw a pipe or other threaded connection tight by the use of a wrench.

MAKE-UP TORQUE
The power necessary to screw a joint of pipe into another sufficiently tight to hold and not loosen under working conditions.

MAKING HOLE
Progress in drilling a well, literally.

MALE CONNECTION
A pipe, rod, or coupling with threads on the outside circumference.

MANHOLE
A hole in the side of a tank or other vessel through which a man can enter. Manholes have fitted covers with gaskets that are kept bolted in place when the tank is in use.

MANIFEST
A document issued by a shipper covering oil or products to be transported by truck.

MANIFOLD
An area where pipelines entering and leaving a pumping station or tank farm converge and that contains all valves for controlling the incoming and outgoing streams.

MAN RACK
A portable "dog house" or cab mounted on a flat-bed truck for transporting pipeline workers to and from the job.

MAP, BASE
A map that contains latitude and longitude lines, land and political boundaries, rivers, lakes, and major cities.

MAP, RELIEF
A model of an area in which variation in the surface is shown in relief; a three-dimensional model of a surface area.

MAP, SURVEY
A map containing geologic information of the surface and/or the subsurface.

MAP, TOPOGRAPHIC
A map which shows in detail the physical features of an area of land, including rivers, lakes, streams, roads.

MARGINAL STRIKE
A discovery well on the border line between what is considered a commercial and a non-commercial well; a step-out well that may have overreached the pool boundary.

MARGINAL WELL
A low-producing well, usually not subject to allowable regulations.

MARINE OIL
Petroleum found by wells offshore or on the continental shelf.

MARINE RISER SYSTEM
A string of specially designed steel pipe that extends down from a drill ship or floating platform to the subsea wellhead. Marine risers are used to provide a return fluid-flow conductor between the well bore and the drill vessel and to guide the drill string to the wellhead on the ocean floor. The riser is made up of several sections including flexible joints and a telescoping joint to absorb the vertical motion of the ship caused by wave action.

MARINE WHITE GASOLINE
Gasoline made for camp stoves, lanterns, blow torches, boat motors. Marine white contains no tetraethyl lead or other additives that could clog the needle valves of gasoline appliances.

MARSH BUGGY
A tractor-like vehicle whose wheels are fitted with extra large rubber tires inflated with air for use in swamps. The great, balloon-like tires are 10 or 12 feet high and two or three feet wide providing buoyancy as well as traction in marshland. The marsh buggy is indispensable in exploration work in swampy terrain.

MASS-FLOW GAS METER
A gas meter that registers the quantity of gas in pounds which is then converted to cubic feet. Mass-flow meters, which are somewhat more accurate than orofice meters, are used in many refineries where large volumes of gas are consumed.

MAST
A simple derrick made of timbers or pipe held upright by guywires; a sturdy A-frame used for drilling shallow wells or for workover; a ginpole.

MASTER GATE
A large valve on the wellhead used to shut in a well if it should become necessary.

MAT STRUCTURE
The steel platform placed on the sea floor as a rigid foundation to support the legs of a jack-up drilling platform.

MAT-SUPPORTED DRILLING PLATFORM
A self-elevating (jack-up) offshore drilling platform whose legs are attached to a metal mat or substructure that rests on the sea floor when the legs are extended.

MATTOCK
A tool for digging in hard earth or rock. The head has two sharpened steel blades; one is in the shape of a pick, the other the shape of a heavy adz.

MAXIMUM EFFICIENT RATE (MER)
Taking crude oil and natural gas from a field at a rate consistent with "good production practice", i.e. maintaining reservoir pressure, controlling water, etc; also the rate of production from a field established by a state regulatory agency.

MCF AND MMCF
Thousand cubic feet; the standard unit for measuring volumes of natural gas. MMCF is one million cubic feet.

MEASURE, UNITS OF

LENGTH

1 Centimeter	= 0.3937 inches	= 0.0328 feet
1 Meter	= 39.37 inches	= 1.0936 yards
1 Kilometer	= 0.6213 miles	= 3,280 feet
1 Foot	= 0.3048 meters	
1 Inch	= 2.54 centimeters	
1 Mil	= 0.001 inch	

SQUARE MEASURE

1 Sq. Centimeter =	0.1550 sq. inches	
1 Sq. Meter =	1.196 sq. yards	= 10.784 sq. feet
1 Sq. Kilometer =	0.386 sq. miles	
1 Sq. Foot	= 929.03 sq. centimeters	
1 Sq. Mile =	2.59 sq. kilometers	
1 Sq. Inch	= 1 million sq. mils	

MER
Maximum efficient rate (of production) (q.v.).

MERCAPTANS
Chemical compounds, containing sulfur, present in certain refined products that impart objectionable odor to the product.

MERCHANTABLE OIL
Oil (crude) of a quality as to be acceptable by a pipeline system or other purchaser; crude oil containing no more than one percent BS&W (q.v.).

MERCURY NUMBER
A measure of the free sulfur in a sample of naphtha. Mercury is mixed with a sample and shaken, and the degree of discoloration in the sample is compared with a standard to determine the mercury number.

METAMORPHISM
Changes in rock induced by pressure, heat, and the action of water that results in a more compact and highly crystalline condition.

METHANE
The simplest saturated hydrocarbon; a colorless flammable gas; one of the main constituents of illuminating gas.

METHANOL
Methyl alcohol; a colorless, flammable liquid derived from methane (natural gas).

METRIC SYSTEM CONVERSION

Inches	x	0.0254	= meters
Feet	x	0.305	= meters
Miles	x	1609.	= meters
Miles	x	1.609	= kilometers
Millimeters	x	0.03937	= inches
Centimeters	x	0.3937	= inches
Meters	x	39.37	= inches
Meters	x	3.281	= feet
Kilometers	x	0.621	= miles
Sq. Centimeters	x	0.155	= sq. inches
Sq. Meters	x	10.764	= sq. feet
Cu. Centimeters	x	0.061	= cu. inches
Liters	x	0.2642	= gallons
Gallons	x	3.78	= liters

METRIC SYSTEM PREFIXES

Micro	= one millionth	Hecto	= one hundred
Milli	= one one-thousandth	Kilo	= one thousand
Centi	= one hundredth	Myria	= ten thousand
Deci	= one tenth	Mega	= one million
Deca	= ten		

MICELLAR/SURFACTANT FLOODING
A tertiary recovery technique; a method of recovering additional crude oil from a field depleted by conventional means including repressuring and water flooding. Micellar-surfactant drive or flooding involves injecting water mixed with certain chemicals into the producing formation. The chemical solution reduces the surface tension of the oil clinging or adhering to the porous rock thus "setting the oil free" to be pumped out with the flooding solution. Such a project may have various names, e.g. micellar; micellar-polymer; soluable-oil; petroleum sulfonate.

MICROBALLOONS
A foam blanket that floats on the liquid in storage tanks to reduce losses from evaporation. The blanket is composed of billions of hollow, balloon-like plastic spheres containing a sealed-in gas—usually nitrogen. The spheres are almost microscopic in size. When poured in sufficient quantity on top of crude oil or refined products in a tank, they spread across the surface forming a dense layer that is effective in reducing evaporation.

MID-CONTINENT CRUDE
Oil produced principally in Kansas, Oklahoma, and North Texas.

MIDDLE DISTILLATES
The term applied to hydrocarbons in the so-called middle range of refinery distillation; e.g. kerosine, light diesel oil, heating oil, and heavy diesel oil.

MIDNIGHT REQUISITION
Obtaining material without proper authority; borrowing unbeknown to the "lender"; swiping for a "good" cause.

MILL
To grind up; to pulverize with a milling tool (q.v.).

MILLABLE
Said of material used down hole, i.e., packers, bridges, and plugs, "soft" enough to be bored out or pulverized with a milling tool.

MILLING
Cutting a "window" in a well's casing with a tool lowered into the hole on the drill string.

MILLING TOOL
A grinding or cutting tool used on the end of the drill column to pulverize a piece of down-hole equipment or to cut the casing.

MILL SCALE
A thin layer or incrustation of oxide which forms on the surface of iron and steel when it is heated during processing. Pipelines must be cleaned of mill scale before being put in service carrying crude oil, gas or products. This is done by running steel-bristle pigs and scrapers.

MINERAL SPIRITS
Common term for naphthas (solvents), those used for dry cleaning and paint thinners.

MINI-SEMI
A scaled-down semisubmersible drilling platform built for service in relatively shallow water.

MMBTU/HR
Million BTU (British thermal units) per hour; rating used for large industrial heaters and other large thermal installations such as furnaces and boilers.

MOBILE PLATFORM
A self-contained, offshore drilling platform with the means for self-propulsion. Some of the larger semi-submersible drilling platforms are capable of moving in the open sea at five to seven knots.

MOCK-UP
A full-sized structural model built accurately to scale for study and testing of an installation to be used or operated commercially. For deep-water, offshore work mock-ups are made to simulate conditions in subsea well-head chambers and sea floor work areas.

MODULE
An assembly (q.v.) that is functional as a unit, and can be joined with other units for increasing or enlarging the function; for example, a gas-compressor module; an electronic or hydraulic module.

MONEY LEFT ON THE TABLE
A phrase referring to the difference between the high and the second highest bid made by operators or companies when bidding on Federal or state oil leases. For example: high bid, $1,000,000; second-highest bid, $750,000. Money left on the table, $250,000.

MONKEY BOARD
A colloquial and humorous reference to the tubing board (q.v.) high in the derrick.

MONKEY WRENCH
An adjustable, square-jawed wrench whose adjusting screw-collar is located on the handle, and whose head can be used as a hammer; a crude wrench suitable for mechanical work of the roughest kind.

MONOCLINE
A geological term for rock strata that dip in one direction. When the crest of an anticline (q.v.) is eroded away, a partial cross section of the strata making up the fold is exposed at the earth's surface and the undisturbed lower flanks form what are called monoclines.

MONOPOD PLATFORM
A type of offshore drilling platform with a single supporting leg. The design of the monopod makes it effective in arctic regions where thick, moving bodies of ice present serious problems for more conventional platforms.

MOONPOOL
The opening in a drill ship through which drilling operations are carried on;

the moonpool or drillwell is usually located amidship, with the derrick rising above.

MOPE POLE
A lever; a pry-pole usually made by cutting a small tree; used on pipeline construction as an adjunct to the jack board and in lowering the pipeline into the ditch.

MORMON BOARD
A broad, reinforced sled-like board with eye bolts on each end and a handle in the center. Used to backfill a pipeline ditch using a team of horses or a tractor pulling the board forward and a workman pulling it back into position for another bite.

MRG PROCESS
Methane Rich Gas Process. MRG is a patented process (Japan Gasoline Co.) to make synthetic natural gas from propane. Liquid propane is hydrodesulfurized and gasified with steam at temperatures between 900° and 1000° F. The resulting gas mixture is methanated, scrubbed to remove CO_2, dried, cooled, and fed to distribution lines.

MUD
See Drilling mud

MUD BARREL
A small bailer used to retrieve cuttings from the bottom of a cable tool drilling well.

MUD-COOLING TOWER
In drilling in or near a geothermal reservoir, the drilling mud becomes superheated and must be cooled to avoid flashing or vaporizing of the liquid (water or oil) in the mud stream at the surface. Cooling also reduces the thermal stress on the drill string.

MUD CUP
A device for measuring drilling mud density or weight; a funnel-shaped cup into which a measured quantity of mud is poured and allowed to run through, against time.

MUD HOG
A mud pump; a pump to circulate drilling mud in rotary drilling; slush pump.

MUD HOSE
The flexible, steel-reinforced, rubber hose connecting the mud pump with the swivel and kelly joint on a drilling rig. Mud is pumped through the mud hose to the swivel and down through the kelly joint and drill pipe to the bottom of the well.

MUD LINE
The sea or lake bottom; the interface between a body of water and the earth.

MUD LOG
A progressive analysis of the well-bore cuttings washed up from the bore hole by the drilling mud. Rock chips are retrieved with the aid of the shale shaker (q.v.) and examined by the geologist.

MUD PITS
Excavations near the rig into which drilling mud is circulated. Mud pumps withdraw the mud from one end of a pit as the circulated mud, bearing rock chips from the bore hole, flows in at the other end. As the mud moves to the suction line, the cuttings drop out leaving the mud "clean" and ready for another trip to the bottom of the bore hole. See Reserve pit.

MUDSCOW
A portable drilling-mud tank in the shape of a small barge or scow used in cable-tool drilling when relatively small amounts of mud were needed or in a location when a mud pit was not practical. Also, a conveyance, a kind of large sled for transporting pipe and equipment into a marshy location. The mudscow is pulled by a crawler-type tractor which would not bog down as would a wheeled vehicle.

MUD TANKS
Portable metal tanks to hold drilling mud. Mud tanks are used where it is impractical to dig mud pits (q.v.) at the well site.

MUD UP, TO
In the early days of rotary drilling and before the advent of accurate well logging, producible formations could be mudded up (plastered over) by the sheer weight of the column of drilling mud, so said the cable-tool men who were skeptical of the new-fangled drilling method. Mudding up occurs also in pumping wells. The mud may be from shaley portions of the producing formation, from sections of uncased hole, or the residue of drilling mud.

MULE SKINNER
Forerunner to the truck driver; a driver of a team or span of horses or mules hitched to an oil field wagon. Unhitched from the wagon, the team was used to pull, hoist, and do earthwork with a slip or Fresno (q.v.). The "skinner" got his name from the ability to skin the hair off a mule's rump with a crack of the long reins he used, appropriately called butt lines.

MULLET
Humorous and patronizing reference to an investor with money to put into the drilling of an oil well with the expectation of getting rich; a sucker; a

person who knows nothing about the oil business or the operator with whom he proposes to deal.

MULTI-BUOY MOORING SYSTEM
A tanker loading facility with five or seven mooring buoys to which the vessel is moored as it takes on cargo or bunkers (q.v.) from submerged hoses that are lifted from the sea bottom. Submarine pipelines connect the pipeline-end manifold to the shore.

MULTI-PAY WELL
See Multiple completion

MULTIPLE COMPLETION
The completion of a well in more than one producing formation. Each production zone will have its own tubing installed, extending up to the Christmas tree. From there the oil may be piped to separate tankage. *See* Dual completion.

MULTIPLIER
A device or linkage for increasing (or decreasing) the length of the stroke or travel of a rod line furnishing power for pumping wells on a lease. A beam which oscillates on a fulcrum-and-bearing to which is attached the rod line from the power source (central power, q.v.) and a rod line to the pumping well. By varying the distance from the fulcrum of the two rod-line connections, the travel of the well's rod line can be lengthened or shortened to match the stroke of the well's pump.

N

NAPHTHA
A volatile, colorless liquid obtained from petroleum distillation; used as a solvent in the manufacture of paint, as dry-cleaning fluid, and for blending with casinghead gasoline in producing motor gasoline.

NATIVE GAS
Gas originally in place in an underground formation as opposed to gas injected into the structure.

NATURAL GAMMA-RAY LOGGING
A procedure in which gamma rays naturally given off or emitted by rock formations, cut through by the well's bore hole, are measured. A radiation detector is lowered into the hole and picks up gamma rays emitted by the

rock. The signals are transmitted to a recording device at the surface. *See* Gamma ray logging.

NATURAL GAS
Gaseous forms of petroleum consisting of mixtures of hydrocarbon gases and vapors, the more important of which are methane, ethane, propane, butane, pentane, and hexane; gas produced from a gas well.

NAVAL PETROLEUM RESERVES
Areas containing proven oil reserves which were set aside for national defense purposes by Congress in 1923. The Reserves, estimated to contain billions of barrels of crude oil, are located in Elk Hills and Buena Vista, California; Teapot Dome, Wyoming, and on the North Slope in Alaska.

NEEDLE VALVE
A valve used on small, high-pressure piping where accurate control of small amounts of liquid or gas is desired. The "tongue" of the valve is a rod that tapers to a point and fits into a seat which permits fine adjustments as when used with pressure gauges.

NEOPRENE
A rubber-like product derived from petroleum and compounded with natural rubber to produce a substance highly resistant to chemicals and oils. Neoprene, first called polychloroprene, was discovered by W. Carothers, Ira Williams, A. Collins, and J. Kirby of the DuPont research laboratory.

NEUTRAL STOCK
Lubricating oil stock that has been dewaxed and impurities removed and can be blended with bright stock (q.v.) to make good lube oil; one of the many fractions of crude oil that, owing to special properties, is ideal as a blending stock for making high-quality lube oil.

N.G.A.
Natural Gas Act. An Act of Congress which empowers the Federal Power Commission to set prices and regulate the transportation of natural gas.

N.G.P.A.
Natural Gas Processors Association, successor to the Natural Gasoline Association of America.

NIPPLE
A short length of pipe with threads on both ends or with weld ends.

NIPPLE CHASER
The material man who serves the drilling rig; the person who makes certain all supplies needed are on hand.

NIPPLE UP
To put together fittings in making a hook up; to assemble a system of pipe, valves, and nipples as in a Christmas tree.

NOBLE-METAL (CATALYST)
A metal used in petroleum refining processes that is chemically inactive with respect to oxygen.

NOISE LOG
A sound detection system inside a logging tool designed to pick up vibrations caused by flowing liquid or gas down hole. The device is used to check the effectiveness of a squeeze job (q.v.), to estimate the gas flow from perforated formations, etc.

NOMINATIONS
(1) The amount of oil a purchaser expects to take from a field as reported to a regulatory agency that has to do with state proration. (2) Information given to the proper agency of the Federal government or a state relative to tracts of offshore acreage a person or company would like to see put up for bid at a lease sale.

NON-OPERATOR
The working-interest owner or owners other than the one designated as operator of the property; a "silent" working-interest owner.

NON-SPARKING TOOLS
Hand tools made of bronze or other non-ferrous alloys for use in areas where flammable oil or gas vapors may be present.

NPR
Naval Petroleum Reserves (q.v.).

N.P.R.A.
National Petroleum Refiners Association

NPT
National pipe thread; denotes standard pipe thread.

"N" STAMP
Designates equipment qualified for use in nuclear installations: pipe, fittings, pumps, valves, etc.

NUT CUTTING, DOWN TO THE
The crucial point; the vital move or decision; a "this is it" situation.

O

OBO VESSEL
A specially designed vessel for carrying ore and crude, both in bulk form. The first oil and bulk ore tanker/carrier was launched in 1966 and used in handling relatively small cargoes of oil and ore.

OCAW
Oil, Chemical and Atomic Workers Union, a labor organization representing a large number of the industry's refinery and other hourly workers.

OCS
Outer Continental Shelf (q.v.).

OCTANE RATING
A performance rating of gasoline in terms of anti-knock qualities. The higher the octane number the greater the anti-knock quality; e.g. 94 octane gasoline is superior in anti-knock qualities to a gasoline of 84 octane.

OD
Outside diameter of pipe; OD and ID (inside diameter) are initials used in specifying pipe sizes; e.g. 4½-inch OD; 8⅝-inch ID.

ODORANT
A chemical compound added to natural gas to produce a detectable, unpleasant odor to alert householders should they have even a small leak in the house piping. Odorants are used also in liquids or gases being stored or transported to detect leaks.

OFF THE SHELF
Said of products or equipment that are ready and waiting at a supplier's warehouse and can be taken "off the shelf" and shipped immediately. Refers also to techniques and procedures that have been perfected and are ready to be employed on some job.

OFF-LOADING
Another name for unloading; off-loading refers more specifically to liquid cargo—crude oil, and refined products.

OFFSET WELL
(1) A well drilled on the next location to the original well. The distance from the first well to the offset well depends upon spacing regulations and whether the original well produces oil or gas. (2) A well drilled on one tract of land to prevent the drainage of oil or gas to an adjoining tract where a well is being drilled or is already producing.

OFFSHORE "WELL NO. 1"
The first offshore well (out of sight of land) was drilled on November 14, 1947, in the Gulf of Mexico, 43 miles south south of Morgan City, Louisiana. By 1976, more than 18,000 wells had been drilled offshore.

OIC
Oil Information Committee of the American Petroleum Institute (API)

OIL
Crude petroleum (oil) and other hydrocarbons produced at the wellhead in liquid form; includes distillates or condensate recovered or extracted from natural gas.

OIL BONUS
A payment in oil to a lessor (usually the landowner) in addition to the cash bonus and royalty payment he is entitled to receive.

OIL COUNTRY TUBULAR GOODS
Well casing, tubing, drill pipe, drill collars, and line pipe.

OILER
The third man at a pumping station in the old days. The normal shift-crew on a large gathering or mainline station was the station engineer, the telegraph operator-assistant engineer, and the oiler whose job included feeling the engine and pump bearings, keeping the wick-oilers full and dripping properly, and cleaning and mopping the station floors.

OIL FINDER
A wry reference to a petroleum geologist.

OIL IMPORT TICKET
A license issued by an agency of the Federal government to refiners to buy certain amounts of crude oil shipped in from abroad.

OIL-IN-PLACE
Crude oil estimated to exist in a field or a reservoir; oil in the formation not yet produced.

OIL-MIST SYSTEM
A lubricating system which pneumatically conveys droplets of a special oil from a central source to the points of application. An oil-mist system is economical in its use of lubricant and efficient on many types of anti-friction applications.

OIL PATCH
A term referring broadly to the oil field, to areas of exploration, production, and pipelining.

OIL PAYMENT
A share of the oil produced from a well or a lease, free of the costs of production.

OIL POOL
An underground reservoir or trap containing oil. A pool is a single, separate reservoir with its own pressure system so that wells drilled in any part of the pool affect the reservoir pressure throughout the pool. An oil field may contain one or more pools.

OIL RING
A metal ring that runs on a horizontal line shaft, in the bearing well which has a supply of lube oil. As the ring slowly rotates through the well of oil, it deposits oil on the shaft. Oil rings are generally made of brass and are used on relatively slow-moving shafts.

OIL ROYALTY
The lessor's or landowner's share of oil produced on his land. The customary ⅛th royalty can be paid in money or in oil. In some instances, another fraction of production is specified as royalty.

OIL RUN
(1) The production of oil during a specified period of time. (2) In pipeline parlance, a tank of oil gauged tested, and put on the line; a run. See Run ticket.

OIL SHALE
Kerogen shale (q.v.).

OIL SPILL
A mishap that permits oil to escape from a tank, an offshore well, an oil tanker, or a pipeline. Oil spill has come to mean oil on a body of water where even small amounts of oil spread and become highly visible.

OIL STRING
See Production string

ON-LINE PLANT (GAS)
Gas processing plant located on or near a gas transmission line which takes gas from the trunk line for processing—stripping, scrubbing, drying— and returns the residue gas to the line.

ON STREAM
Term used for a processing plant, a refinery or pumping station that is operating.

ON THE BEAM
Refers to a well on the pump, operated by a walking beam instead of a pumping jack.

ON THE LINE
(1) Said of a tank of oil whose pipeline valve has been opened and the oil is running into the line. (2) A pumping unit that has been started and is pumping on the pipeline.

ON THE PUMP
A well that is not capable of flowing and is produced by means of a pump.

OOG
Office of Oil and Gas, Department of the Interior

OP-DRILLING SERVICE
Optimization drilling; a consulting service first developed by American Oil Company that makes available to operators of drilling rigs technical, geological, and engineering information gathered from wells drilled in the same area. Included is advice on mud programs, bits, drill speed, pressures, as well as consultation with drilling experts.

OPEC
Organization of Petroleum Exporting Countries (q.v.)

OPEN HOLE
An uncased well bore; the section of the well bore below the casing; a well in which there is no protective string of pipe.

OPERATING INTEREST
An interest in oil and gas that bears the costs of development and operation of the property; the mineral interest less the royalty. *See* Working interest.

OPERATOR
An actuating device; a mechanism for the remote operation and/or control of units of a processing plant. Operators usually are air or hydraulically actuated. Their main use is for opening and closing stops and valves.

OPERATOR, PLANT OR STATION
A worker who is responsible for the operation of a small plant or a unit of a larger plant during his working shift. In the old days, an operator was a telegrapher at a pumping station who sent reports on pumping rates, tank gauges, and line pressures to the dispatcher (q.v.).

ORGANIC SUBSTANCE
A material that is or has been part of a living organism. Oil, although classified as a mineral, is an organic substance derived from living organisms.

ORGANIZATION OF PETROLEUM EXPORTING COUNTRIES
Oil producing and exporting countries in the Middle East, Africa, and

South America that have organized for the purpose of negotiating with oil companies on matters of oil production, prices, future concession rights.

O-RING
A circular rubber gasket used in flanges, valves, and other equipment for making a joint pressure tight. O-rings in cross section are circular and solid.

"OR" LEASE
One of two principal forms of oil and gas leases. The other is the "unless" lease (q.v.). In an OR lease, the lessee promises to drill within a year from the date of the lease *or* pay rental or surrender the lease.

OROFICE METER
A measuring instrument that records the flow rate of gas, enabling the volume of gas delivered or produced to be computed.

O&S
Over and short (q.v.)

OSHA
Occupational Safety & Health Administration

OTTO-CYCLE ENGINE
A four-stroke cycle gas engine; the conventional automobile engine is an Otto-cycle engine, invented in 1862 by Beau de Rochas and applied by Dr. Otto in 1877 as the first commercially successful internal combustion engine. The four strokes of the Otto cycle are: intake, compression, power, and exhaust.

OUTAGE GAUGE
A measure of the oil in a tank by finding the distance between the top of the oil and the top of the tank and subtracting this measurement from the tank height.

OUTCROP
A subsurface rock layer or formation that, owing to geological conditions, appears on the surface in certain locations. That part of a strata of rock that comes to the surface.

OUTER CONTINENTAL SHELF (OCS)
"All submerged lands (1) which lie seaward and outside the area of lands beneath the navigable waters as defined in the Submerged Lands Act (67 Stat. 29) and (2) of which the subsoil and seabed appertain to the U.S. and are subject to its jurisdiction and control."

OUTPOST WELL
A well drilled in the hope of making a long extension to a partly developed

field; an ambitious, giant-stride, step-out well that hopes to pick up the same pay zone as that encountered in the field.

OVER AND SHORT (O&S)
In a pipeline gathering system O&S refers to the perennial imbalance between calculated oil on hand and the actual oil on hand. This is owing to contraction, evaporation, improper measuring of lease tanks, and losses through undetected leaks. Oil is paid for on the basis of the amount shown in the lease tanks. By the time this oil is received at the central gathering station, the amounts invariably are short which represent a loss to the pipeline system.

OVERHEAD
A product or products taken from a processing unit in the form of a vapor or a gas; a product of a distillation column.

OVER-RIDE
See Overriding royalty

OVERRIDING ROYALTY
An interest in oil and gas produced at the surface free of any cost of production; royalty in addition to the usual landowner's royalty reserved to the lessor. A 1/16th over-ride is not unusual.

OVERRIDE SYSTEM
A backup system; controls that take over should the primary system of controls fail or are taken out for adjustment or repair; a redundancy built in for safety and operational efficiency.

OVERSHOT
A fishing tool; a specially designed barrel with gripping lugs on the inside that can be slipped over the end of tubing or drill pipe lost in the hole. An overshot tool is screwed to a string of drill pipe and lowered into the hole, and over the upper end of the lost pipe. The lugs take a friction grip on the pipe which can then be retrieved.

OVER-THE-DITCH COATING
Doping and wrapping line pipe above the ditch just before it is lowered in. Most line pipe is coated and wrapped in the pipe yard and then trans-ported to the right-of-way and strung. Over-the-ditch coating has the advantage of minimizing scuffing or other damage to the coating suffered through moving and handling.

OXYACETYLENE WELDING
The use of a mixture of oxygen and acetylene in heating and joining two pieces of metal. When the weld edges of the two pieces are molten, metal from a welding rod is melted onto the molten puddle as the welder holds

the tip of the rod in the flame of the torch. Oxygen and acetylene are used also in cutting through metal. The intense heat generated at the tip of the cutting torch (about 3,500° F) literally melts away the metal in the area touched by the flame. *See* Welding torch.

P

PACKAGE PLANT
A facility at a refinery where various refined products are put in cartons and boxes ready for shipment. Waxes, greases, and small-volume specialty oils are boxed in a package plant.

PACKED-HOLE ASSEMBLY
A drill column containing special tools to stabilize the bit and keep it on a vertical course as it drills. Included among the tools are stabilizer sleeves (q.v.), square drill collars (q.v.), and reamers. Packed-hole assemblies are often used in "crooked-hole country."

PACKER
An expanding plug used in a well to seal off certain sections of the tubing or casing when cementing, acidizing, or when a production formation is to be isolated. Packers are run on the tubing or the casing, and when in position can be expanded mechanically or hydraulically against the pipe wall or the wall of the well bore.

PACKING
Any tough, pliable material—rubber or fiber—used to fill a chamber or "gland" around a moving rod or valve stem to prevent the escape of gas or liquid; any yielding material used to effect a pressure-tight joint. Packing is held in place and compressed against a moving part by a "follower," an adjustable element of the packing gland.

PACKING GLAND
A stuffing box; a chamber that holds packing material firmly around or against a moving rod or valve stem to prevent the escape of gas or liquid. An adjustable piece that fits into the gland to compress the packing against the moving part is called the "follower" and can be screwed into the gland or forced into the gland by nuts on stud bolts.

PARAFFIN
A white, odorless, tasteless, and chemically inert waxy substance derived

from distilling petroleum; a crystalline, flammable substance composed of saturated hydrocarbons.

PARAFFIN-BASE CRUDE
Crude oil containing little or no asphalt materials; a good source of paraffin, quality motor lubricating oil, and high-grade kerosine; usually has lower non-hydrocarbon content than an asphalt-base crude.

PARTICIPATION
A type of joint venture between a host country and an international oil company holding concession rights in that country. Participation may be voluntary on the part of the oil company or as the result of coercion by the host country.

PARTICULATE MATTER
Minute particles of solid matter—cinders and fly-ash—contained in stack gases.

PAY HORIZON
The subsurface, geological formation where a deposit of oil or gas is found in commercial quantities.

PAY OUT
The recovery from production of the costs of drilling, completion, and equipping a well. Sometimes included in the costs is a prorata share of lease costs.

PAY STRING
The pipe through which a well is produced from the pay zone. Also called the "long string" because only the pay string of pipe reaches from the wellhead to the producing zone.

PCV
Positive Crankcase Ventilation (q.v.)

PEAK-SHAVING LNG PLANT
A liquefied natural gas plant that supplies gas to a gas pipeline system during peak-use periods. During slack periods the liquefied gas is stored. With the need for additional gas, the liquid product is gasified and fed into the gas pipeline.

PEA PICKER
An inexperienced worker; a green hand; boll weevil.

PENDULUM DRILL-ASSEMBLY
A heavily weighted drill-assembly using long drill collars and stabilizers to help control the drift from the vertical of the drill bit. The rationale for the

weighted drill assembly is that, like a pendulum at rest, it will resist being moved from the vertical and will tend to drill a straighter hole.

PENNSYLVANIA-GRADE CRUDE OIL
Oil with characteristics similar to the crude oil produced in Pennsylvania from which superior quality lubricating oils are made. Similar-grade crude oil is also found in West Virginia, eastern Ohio, and southern New York state.

PERCENTAGE DEPLETION
A method of computing the allowance for depletion of an oil or gas well, or other mining of minerals, for Federal income tax purposes. A provision in the tax law that exempts a certain percent of mineral production from income tax. For an oil well, the percentage depletion rate is 22 percent of the well's gross production, excluding royalty, up to 50 percent of the net income from the property.

PERFORATING
To make holes through the casing opposite the producing formation to allow the oil or gas to flow into the well. Shooting steel bullets through the casing walls with a special downhole "gun" is a common method of perforating.

PERMEABILITY
A measure of the resistance offered by rock to the movement of fluids through it. Permeability is one of the important properties of sedimentary rock containing petroleum deposits. The oil contained in the pores of the rock cannot flow into the well bore if the rock in the formation lacks sufficient permeability. Such a formation is referred to as "tight." See Porosity.

PERSUADER
An oversize tool for a small job; an extension added to the handle of a wrench to increase the leverage.

PESA
Petroleum Equipment Suppliers Association

PETROCHEMICALS
Chemicals derived from petroleum; feedstocks for the manufacture of a variety of plastics and synthetic rubber.

PETROCHEMISTRY
A word derived from petroleum and chemistry; the science of synthesizing substances derived from crude oil, natural gas, and natural gas liquids.

PETROFRACTURING
A process in which a mixture of oil, chemicals, and sand is pumped under

high pressure into an oil-bearing formation penetrated by the well bore. This produces cracks and fissures in the formation to improve the flow of oil. *See* Hydraulic fracturing.

PETROLEUM
In its broadest sense, the term embraces the whole spectrum of hydrocarbons—gaseous, liquid, and solid. In the popular sense, petroleum means crude oil.

PETROLEUM ROCK
Sandstone, limestone, dolomite, shale, and other porous rock formations where accumulations of oil and gas may be found.

PETROLEUM TAR SANDS
Native asphalt, solid and semi-solid bitumen, including oil-impregnated rock or sands from which oil is recoverable by special treatment. Processes have been developed for extracting the oil, referred to as synthetic crude.

PH (pH)
A symbol used in expressing both acidity and alkalinity on a scale whose values run from 0 to 14, with 7 representing neutrality; numbers less than 7, increasing acidity; greater than 7, increasing alkalinity.

PHOTOMETRIC ANALYZER
A device for detecting and analyzing the changes in properties and quantities of a plant's stack gases. The analyzer, through the use of electronic linkage, automatically sounds a warning or effects changes in the stack emissions.

PHYSICAL DEPLETION
The exhausting of a mineral deposit or a petroleum reservoir by extraction or production.

PICKLE
A cylindrical weight (two to four feet in length) attached to the end of a hoisting cable, just above the hook, for the purpose of keeping the cable hanging straight and thus more manageable for the person using the wire line.

PIER
A walkway-like structure built on piling out from shore, a distance over the water for use as a landing place or to tie up boats.

PIG
A cylindrical device (three to seven feet long) inserted in a pipeline for the purpose of sweeping the line clean of water, rust, or other foreign matter. When inserted in the line at a "trap", the pressure of the oil stream behind it

pushes the pig along the line. Pigs or scrapers (q.v.) are made with tough, pliable discs that fit the internal diameter of the pipe, thus forming a tight seal as it moves along cleaning the pipe walls.

PIG TRAP
A scraper trap (q.v.).

PILELESS PLATFORM
A concrete offshore drilling platform of sufficient weight to hold the structure firmly in position on the sea bottom. Referred to as a "gravity structure", the platform is constructed on shore and then floated and towed to location where it is "sunk" by flooding its compartments. Some platforms of this type have oil storage facilities within the base of the structure. See Gravity structure; also Tension-leg platform.

PILOT PLANT
A small model of a future processing plant, used to develop and test processes and operating techniques before investing in a full-scale plant.

PINCHING A VALVE
Closing a valve part way to reduce the flow of liquid or gas through a pipeline. See Cracking a valve.

PINCH-OUT
The disappearance or "wedging out" of a porous, permeable formation between two layers of impervious rock. The gradual, vertical "thinning" of a formation, over a horizontal or near-horizontal distance, until it disappears.

PIPE FITTER
One who installs and repairs piping, usually of small diameter. An "oil patch plumber" according to pipeliners who traditionally work with large diameter pipe.

PIPE FITTINGS
See Fittings

PIPELINE CAT
A tough, experienced pipeline construction worker who stays on the job until it is flanged up and then disappears—until the next pipeline job. A hard-working, impermanent construction hand; a boomer.

PIPELINE GAS
Gas under sufficient pressure to enter the high-pressure gas lines of a purchaser; gas sufficiently dry so that liquid hydrocarbons—natural gasoline, butane, and other gas liquids, usually present in natural gas—will not condense or drop out in the transmission lines.

PIPELINE GAUGER
See Gauger

PIPELINE OIL
Clean oil; oil free of water and other impurities so as to be acceptable by a pipeline system.

PIPELINE PATROL
The inspection of a pipeline for leaks, washouts, and other unusual conditions by the use of light, low-flying aircraft. The pilot reports by radio to ground stations on any unusual condition on the line.

PIPELINE PRORATIONING
The refusal by a purchasing company or a pipeline to take more oil than it needs from the producer, by limiting pipeline runs from the producer's lease; an informal practice in the days of over-production, when market conditions were unsatisfactory, or when the pipeline system lacked storage space. Also referred to as Purchaser prorationing.

PIPELINE RIDER
One who covers a pipeline by horseback looking for leaks in the line or washed out sections of the right-of-way. The line rider has been replaced by the pipeline patrol using light planes or, for short local lines, by the pickup truck and the man on foot.

PIPELINE WELDING
In pipeline welding, the bevelled ends of two joints are brought together and aligned by clamps. Welders then lay on courses of weld-metal called passes or beads designated as: (1) stringer bead, (2) hot pass, (3) third pass or hot fill (for heavy-wall pipe), (4) filler pass, and (5) final or capping pass.

PIPE SLING
A stirrup-like sling made of heavy belting material used on the winch line of boom-cats for lifting, handling, and lowering in of pipe. Fabric slings are used to prevent scarring or damaging the pipeline's protective coating.

PIPE STRAIGHTENER
A heavy, pipe-yard press equipped with hydraulically powered mandrels for taking the kinks and bends out of pipe. The replaceable mandrels come in all sizes, 2″ to 12″.

PIPE TONGS
Long-handled wrenches that grip the pipe with a scissor-like action; used in laying a screw pipeline. The head (called the butt) is shaped like a parrot's beak and uses one corner of a square "tong key", held in a slot in the head, to bite into the surface of the pipe in turning it.

PITCH
Asphalt; a dark brown to black bituminous material found in natural beds, and is also produced as a black, heavy residue in oil refining. *See* Brea.

PITCHER PUMP
A small hand-pump for very shallow water wells. Looking much like a large, cast-iron cream pitcher, the pitcher pump is built on the order of the "old town pump" with one exception. The pitcher pump's handle, working on a fulcrum, does not have a string of pump rods attached. Water is pumped by the suction created by a leather cup and valve arrangement in the throat or lower body of the pump together with a foot valve 20 feet or so down in the tubing. A simple and elegantly fundamental pumping machine.

PIT LINERS
Specially-formulated plastic sheeting for lining earthen or leaking concrete pits to prevent seepage of oil or water into the ground.

PITMAN
The connecting piece between the crank on a shaft and another working part. On cable tool rigs, the pitman transmits the power from the band-wheel crank to the walking beam (q.v.).

PITOT TUBE
A measuring device for determining the gas-flow rates during tests. The device consists of a tube with a ⅛th-inch inside diameter inserted in a gas line horizontal to the line's long axis. The impact pressure of the gas flow at the end of the tube compared to the static pressure in the stream are used in determining the flow rate.

PITTED PIPE
Line pipe corroded in such a manner as to cause the surface to be covered with minute, crater-like holes or pits.

PLANT OPERATOR
An employee who runs plant equipment, makes minor adjustments and repairs, and keeps the necessary operating records.

PLASTIC FLOW
The flow of liquid (through a pipeline) in which the liquid moves as a column; flowing as a river with the center of the stream moving at a greater rate than the edges which are retarded by the friction of the banks (or pipe wall). *See* Turbulent flow.

PLAT
A map of land plots laid out according to surveys made by Government Land Office showing section, township, and range; a grid-like representa-

tion of land areas showing their relationship to other areas in a state or county.

PLATT BOOK
A book containing maps of land plots arranged according to Township and Range for counties within a state. *See* Plat.

PLATFORM BURNER
See Forced-draft burner

PLATFORM JACKET
A supporting structure for an offshore platform consisting of large-diameter pipe welded together with pipe braces to form a four-legged stool-like structure (stool without a seat). The jacket is secured to the ocean floor with piling driven through the legs. The four-legged offshore platform is then slipped into legs of the jacket and secured with pins and by the weight of platform and equipment.

PLATFORMATE
High-octane gasoline blending stock produced in a catalytic reforming unit, commonly known as a platformer (q.v.).

PLATFORMER
A catalytic reforming unit which converts low-quality straight-chain paraffins or naphthenes to low-boiling, branched-chain paraffins or aromatics of higher octane; a refinery unit that produces high-octane blending stock for the manufacture of gasoline.

PLEM
Pipeline-end manifold; an offshore, submerged manifold connected to the shore by pipelines that serve a tanker loading station of the multi-buoy mooring type (q.v.).

PLENUM
A room or enclosed area where the atmosphere is maintained at a pressure greater than the outside air. Central control rooms at refineries are usually kept at pressures of a few ounces above the surrounding atmosphere to prevent potentially explosive gases from seeping into the building and being ignited by electrical equipment. Some offshore drilling and production platforms are provided with plenums as a safety measure. *See* Acoustic plenum.

PLUG
To fill a well's bore hole with cement or other impervious material to prevent the flow of water, gas, or oil from one strata to another when a well is abandoned; to screw a metal plug into a pipeline to shut off drainage or to divert the stream of oil to a connecting line; to stop the flow of oil or gas.

PLUG BACK
To fill up the lower section of the well bore to produce from a formation higher up. If the well has been cased, the casing is plugged back with cement to a likely formation and then perforated (q.v.). *See* Bridge plug.

PLUGGING A WELL
To fill up the bore hole of an abandoned well with mud and cement to prevent the flow of water or oil from one strata to another or to the surface. In the industry's early years, wells were often improperly plugged or left open. Modern practice requires that an abandoned well be properly and securely plugged.

PLUG VALVE
A type of quick-opening pipeline valve constructed with a central core or "plug." The valve can be opened or closed with one-quarter turn of the plug; a stop.

PLUNGER
The piston in the fluid end of a reciprocating pump. *See* Plunger pump.

PLUNGER PUMP
A reciprocating pump in which plungers or pistons moving forward and backward or up and down in cylinders drawing in a volume of liquid and, as a valve closes, pushes the fluid out into a discharge line.

POINT MAN
The member of a pipeline tong crew who handles the tips (the points) of heavy pipe-laying tongs. He is the "brains" of the crew as he keeps his men pulling and "hitting" in unison and in time with the other tong crews working on the same joint of screw pipe.

POLISHED ROD
A smooth brass or steel rod that works through the stuffing box or packing gland of a pumping well; the uppermost section of the string of sucker rods, attached to the walking beam of the pumping jack.

POLYETHYLENE
A petroleum-derived plastic material used for packaging, plastic housewares, and toys. The main ingredient of polyethylene is the petrochemical gas ethylene.

POLYMERIZATION
A refining process of combining two or more molecules to form a single heavier molecule; the union of light olefins to form hydrocarbons of higher molecular weight; polymerization is used to produce high-octane gasoline blending stock from cracked gases.

PONTOONS
The elements of a floating roof tank that provide buoyance; air-tight, metal tanks that float on the fluid and support the moveable deck structure of the roof.

PONY RODS
Sucker rod made in short lengths of 2' to 8' for use in making up a string of pumping rods of the correct length to connect to the polished rod of the pumping jack. Pony rods are screwed into the top of the string just below the polished rod (q.v.).

POOL
See Oil pool

POP-OFF VALVE
See Relief valve

POPPING
The discharge of natural gas into the atmosphere; a common practice in the 1920s and 1930s, especially with respect to sour gas and casinghead gas. After the liquid hydrocarbons were extracted, the gas was "wasted" as there was no ready market for it.

PORCUPINE
A cylindrical steel drum with steel bristles protruding from the surface; a super, pipe-cleaning pig for swabbing a sediment-laden pipeline.

POROSITY
The state or quality of being porous; the volume of the pore space expressed as a percent of the total volume of the rock mass; an important property of oil-bearing formations. Good porosity indicates an ability to hold large amounts of oil in the rock. And with good permeability (q.v.), the quality of a rock that allows liquids to flow through it readily, a well penetrating the formation should be a producer.

PORTABLE PIPE MILL
A very large, self-propelled, "factory-on-wheels" that forms, welds, and lays line pipe in one continuous operation. The pipe is made from rolls of sheet steel (skelp) shaped into a cylindrical form, electric welded, tested, and strung out behind the machine as it moves forward.

POSITIVE CRANKCASE VENTILATION SYSTEM
A system installed on automobiles manufactured after 1968 to reduce emissions from the engine's crankcase. The emissions—oil and un-burned gasoline vapors—are directed into the intake manifold and from there they mix with the gasoline to be burned.

POSITIVE-DISPLACEMENT PUMP
A pump that displaces or moves a measured volume of liquid on each stroke or revolution; a pump with no significant slippage; a plunger or rotary pump.

POSSUM BELLY
A metal box built underneath a truck bed to hold pipeline repair tools—shovels, bars, tongs, chains, and wrenches.

POSTED PRICE
The price an oil purchaser will pay for crude oil of a certain API gravity and from a particular field or area. Once literally posted in the field, the announced price is now published in area newspapers. With government control affecting almost all aspects of the industry, prices of oil and gas are not permitted to be set by the industry's supply and demand requirements as they once were. *See* FEA.

POTS (PUMP VALVE)
See Valve pots

POUR POINT
The temperature at which a liquid ceases to flow; or at which it congeals.

POUR-POINT DEPRESSANT
A chemical agent added to oil to keep it flowing a low temperatures.

POWER
See Central power

POWER TAKEOFF
A wheel, hub, or sheave that derives its power from a shaft or other driving mechanism connected to an engine or electric motor; the end of a power shaft designed to take a pulley.

POWER TONGS
An air or hydraulically-powered mechanism for making up and breaking out joints of drill pipe, casing or tubing. After a joint is stabbed, the power tongs are latched onto the pipe which is screwed in and tightened to a predetermined torque.

PPM
Parts per million; a measure of the concentration of foreign matter in air or a liquid.

PPM/VOL
Parts per million (of water) in a given volume of natural gas. *See* also LBS-H_2O/MMSCF.

PPM/WT
Parts per million (of water) in a given weight of gas; used to express water content in a small amount of gas. *See* also LBS-H$_2$O/MMSCF.

PRAIRIE-DOG PLANT
A small, basic refinery located in a remote area.

PRESSURE MAINTENANCE
See Repressuring operation

PRESSURE SNUBBER
A pulsation dampener (q.v.).

PRIME MOVER
The term describes any source of motion; in the oil field it refers to engines and electric motors; the power source. Prime mover is also applied to large four-wheel-drive trucks or tractors.

PRIVATE BRAND DEALER
A gasoline dealer who does not buy gasoline from a "major" supplier, but retails under the brand name of an independent supplier or his own brand name.

PRODUCING PLATFORM
An offshore structure with a platform raised above the water to support a number of producing wells. In offshore operations, as many as 60 wells are drilled from a single large platform by slanting the holes at an angle from the vertical away from the platform. When the wells are completed, the drilling equipment is removed and the platform is given over to operation of the producing wells.

PRODUCT GAS
End product gas; gas resulting from a special manufacturing process; synthetic natural gas (q.v.).

PRODUCT IMPORT TICKET
A license issued by an agency of the Federal government to a refiner or marketer to import product from abroad.

PRODUCTION PACKER
An expandable plug-like device for sealing off the annular space between the well's tubing and the casing. The production packer is run as part of the tubing string, inside the casing; when lowered to the proper depth, the packer is mechanically or hydraulically expanded and "set" firmly against the casing wall isolating the production formation from the upper casing while permitting the oil or gas to flow up the tubing.

PRODUCTION PAYMENT LOAN
A loan that is to be repaid out of the production of a well.

PRODUCTION PLATFORM
An offshore structure built for the purpose of providing a central receiving point for oil produced in an area of the offshore. The production platform supports receiving tanks, treaters, separators, and pumping units for moving the oil to shore through a submarine pipeline.

PRODUCTION STRING
The casing set just above or through the producing zone of a well. The production string is the longest and smallest diameter casing run in a well. It reaches from the pay zone to the surface.

PRODUCTION TREE
See Christmas tree

PROPPANTS
Material used in hydraulic fracturing (q.v.) for holding open the cracks made in the formation by the extremely high pressure applied in the treatment; the sand grains, beads, or other miniature pellets suspended in the fracturing fluid that are forced into the formation and remain to prop open the cracks and crevices permitting the oil to flow more freely.

PROPRIETARY DATA
Information on subsurface, geological formations gathered or purchased from a supplier of such data by an operator and kept secret; land and offshore reconnaissance surveys from seismic, magnetic, and gravity studies that are privately owned.

PRORATIONING
Restriction of oil and gas production by a state regulatory commission, usually on the basis of market demand. Prorationing involves allowables which are assigned to fields, and from fields to leases, and then allocated to individual wells.

PROTECTIVE STRING
A string of casing used in very deep wells and run on the inside of the outermost casing to protect against the collapsing of the outer string from high gas pressures encountered.

PROVEN RESERVES
Oil which has been discovered and determined to be recoverable but is still in the ground.

P.S.I.A.
Pounds per square inch absolute; pressure measurement which includes atmospheric pressure.

P.S.I.G.
Pounds per square inch gauge (as observed on a gauge).

PULLED THREADS
Stripped threads; threads on pipe or tubing damaged beyond use by too much torque or force used in making up the joint.

PULLING MACHINE
A pulling unit (q.v.).

PULLING RODS
The operation of removing the pumping or sucker rods from a well in the course of bringing up the bottom-hole pump for repairs or replacement. Rods must also be pulled if they have parted down hole. The rods above the break are pulled in a normal manner; the lower section must first be retrieved with a "fishing tool" (q.v.).

PULLING THE CASING
Removing the casing from the hole after abandoning the well. Prior to plugging the well with mud and cement, as much of the casing as can be pulled is retrieved. It is rare that all the casing can be removed from the hole. Often part of the string must be cut off and left in the hole.

PULLING TOOLS
Taking the drill pipe and bit out of the hole. If the tools are to be run again (put back in the hole), the drill pipe is unscrewed in two or three-joint sections (stands) and stacked in the derrick. *See* Doubles.

PULLING UNIT
A portable, truck-mounted mast equipped with winch, wire lines, and sheaves, used for pulling rods or well workover.

PULL ONE GREEN
To pull a drill bit from the hole before it is worn out; to pull a bit before it is necessary.

PULL/ROD LINE
See Shackle rod

PULL RODS
Shackle rods (q.v.).

PULSATION DAMPENER
Various devices for absorbing the transient, rhythmic surges in pressure that occur when fluid is pumped by reciprocating pumps. On such pumps air chambers (q.v.) are installed on discharge lines, which act as air cushions. To protect pressure gauges and other instruments from the incessant pounding, fine-mesh, sieve-like disks are placed in the small tubing or piping to which the gauge is attached; this arrangement "filters out" much of the surging which can damage delicate gauges.

PUMP, CASING
A sucker-rod pump designed to pump oil up through the casing instead of the more common method of pumping through tubing. A casing pump is run into the well on the sucker rods; a packer (q.v.) on top or bottom of the pump barrel provides packoff or seal between the pump and the wall of the casing at any desired depth. Oil is discharged from the pump into the casing and out the wellhead.

PUMP, DOUBLE-DISPLACEMENT
A type of downhole, rod pump which has plungers placed in tandem and operated simultaneously by the pump rods.

PUMP, DUPLEX
A two-cylinder reciprocating plunger pump.

PUMP OFF
To pump a well so rapidly that the oil level falls below the pump's standing valve; to pump a well dry, temporarily.

PUMPER
A person who operates a well's pumping unit, gauges the lease tanks, and keeps records of production; a lease pumper.

PUMPING, BACKSIDE
An arrangement that permits one prime mover (electric motor or engine) to operate two pumping wells. The hook up is such that the down-stroke load on one well counterbalances the upstroke load of the other well. See also Central power.

PUMPING UNIT
A pump connected to a source of power; an oil well pumping jack; a pipeline pump and engine.

PUMPING UNIT, BEAM BALANCED
An oil well pumping unit that carries its well-balancing weights on the walking beam on the end opposite the pump rods. The weights are usually in the form of heavy iron plates added to the walking beam until they balance the pull or weight of the string of pumping rods.

PUMPING UNIT, CRANK BALANCED
An oil well pumping unit that carries its counterweights on the two cranks that flank the unit's gear box. The string of pump rods is balanced by adding sufficient extra iron weights to the heavy cranks. The walking beam on this type unit is short and is not used as a balancing member.

PUMP, JERKER
A single-barrel, small-volume plunger pump actuated by the to-and-fro motion of a shackle-rod line and an attached counterweight. The jerker

pumps on the pull stroke of the rod line; it takes in fluid (the suction stroke) as the counterweight pulls the plunger back from the pumping stroke. Jerkers pump small volumes but can buck high pressure.

PUMPS, ROD

A class of downhole pumps in which the barrel, plunger, and standing valve are assembled and lowered into the well through the tubing. When lowered to its pumping position, the pump is locked to the tubing to permit relative motion between plunger and barrel. The locking device is a hold-down, and consists either of cups or a mechanical, metal-to-metal seal.

PUMP, ROD-LINE

An oil-well pump operated by a shackle-rod line; a pumping jack. *See* Rocker.

PUMP, SIMPLEX

A one-cylinder steam pump used in refineries and processing plants where extra or excess steam is available. Simplex pumps are simple, direct-acting pumps with the steam piston connected directly to the pump's fluid plunger.

PUMP SPECIFICATIONS

A plunger pump designated as 6 x 12 duplex is a two-cylinder pump whose cylinders are 6 inches in diameter with a stroke of 12 inches. A pump with replaceable liners (cylinders) may carry a specifications plate that reads: 4 - 6 x 10. This pump can be fitted with liners and pistons from 4 inches to 6 inches in diameter; its stroke is 10 inches.

PUMP, SUBMERSIBLE

A bottom-hole pump for use in an oil well when a large volume of fluid is to be lifted. Submersible pumps are run by electricity and, as the name implies, operate below the fluid level in the well.

PUMP, TRAVELING-BARREL

A down-hole pump, operated by rods, in which the barrel moves up and down over the plunger, instead of the plunger reciprocating in the barrel as in more conventional pumping devices.

PUMPS, TUBING

A class of downhole pumps in which the barrel of the pump is an integral part of the tubing string. The barrel is installed on the bottom of the string of tubing and is run into the well on the tubing string. The plunger assembly is lowered into the pump barrel on the string of pump rods.

PUP JOINT

A joint of pipe shorter than standard length; any short piece of usable line pipe.

PUT ON THE PUMP
To install a pumping unit on a well. Some wells are pumped from the time they are brought in or completed; others flow for a time (sometimes for many years) and then must be put on the pump.

PVC
Polyvinyl chloride; a commercial resin derived from petroleum; the principal ingredient of PVC is ethylene. The resin can be molded, extruded, or made into a thin, tough film.

PYROBITUMEN
See Kerogen and Kerogen shales

PYROMETER
An instrument for measuring very high temperatures, beyond the range of mecury thermometers. Pyrometers use the generation of electric current in a thermocouple (q.v.) or the intensity of light radiated from an incandescent body to measure temperatures.

Q

QUENCH OIL
A specially refined oil with a high flashpoint (q.v.) used in steel mills to cool hot metal.

QUITCLAIM
An instrument or document releasing a certain interest in land owned by a grantor at the time the agreement takes effect. The key phrase of a quitclaim is: " . . . to release, remise, and forever quitclaim all right, title, and interest in the following described land."

R

RABBIT
A plug put through lease flow-lines for the purpose of clearing the lines of foreign matter, water, and to test for obstructions. *See* Pig.

RACK PRICING
Selling to petroleum jobbers or other resellers at F.O.B. the refinery, with the customers picking up pipeline or other transportation charges.

RACKING BOARD
A platform high in the derrick, on well-service rigs, where the derrick man stands when racking tubing being pulled from the well.

RAFFINATE
In solvent-refining practice, raffinate is that portion of the oil being treated that remains undissolved and is not removed by the selective solvent.

RAINBOW
(1) The irridescence (blues, greens, and reds) imparted to the surface of water by a thin film of crude oil. (2) The only evidence of oil from an unsuccessful well—"Just a rainbow on a bucket of water."

RAM
A closure-mechanism on a blow-out preventer stack; a hydraulically operated type of valve designed to close in a well as with a conventional valve or to close on tubing or drill pipe and maintain high-pressure contact.

RAM, SHEAR
A closure mechanism on a well's blowout-preventer stack fitted with chisel-like jaws that are hydraulically operated. When the ram is closed on the pipe the jaws or blades cut the pipe, permitting the upper section to be removed from the BOP stack.

RANGE OIL
Kerosine-type product used in oil or kerosine stoves or cooking ranges.

RATABLE TAKE
(1) Production of oil and/or gas in such quantities that each landowner whose property overlies a producing formation will be able to recover an equitable share of the oil and/or gas originally in place beneath his land. (2) Production in accordance with allowables set by a state regulatory commission. (3) In some states, common carriers (q.v.) and common purchasers of gas and oil are prohibited from discriminating in favor of one supplier over another.

RATHOLE
(1) A slanted hole drilled near the well's bore hole to hold the kelly joint when not in use. The kelly is unscrewed from the drill string and lowered into the rathole as a pistol into a scabbard. (2) The section of the bore hole that is purposely deviated from the vertical by the use of a whipstock (q.v.).

RATHOLE AHEAD
To drill a hole of reduced diameter in the bottom of the regular bore hole to facilitate the taking of a drill-stem test (q.v.).

RAW GAS
Gas straight from the well before the extraction of the liquefied hydrocarbons (gasoline, butane); wet gas.

RAW-MIX STREAM
A mixture of natural gas liquids being pumped through a pipeline; commingled gas liquids.

R&D
Research and Development; often used to denote a function up to the stage where the commercial potential of a process or technology can be evaluated. *See* Pilot plant.

REAMER
A tool used to enlarge or straighten a bore hole; a milling tool used to cut the casing downhole. Reamers are run on the drill string and are built with cutting blades or wheels that can be expanded against the walls of the hole.

RECLAIMED OIL
Lubricating oil which, after undergoing a period of service, is collected, re-refined, and sold for reuse.

RECTIFIER BED
A source of electric current for protection against corrosion of pipelines, tanks, and other metal installations buried or in contact with the earth. Using a source of AC electric current, the rectifier installation converts the AC to DC (direct current) and allows the DC to flow into the metal to be protected. By reversing the flow of electric current, the corrosion is inhibited. Metal corrosion is a chemical action which produces minute quantities of current that normally flows away from the metal into the ground.

REDUCED CRUDE OIL
Crude oil that has undergone at least one distillation process to separate some of the lighter hydrocarbons. Reducing crude lowers its API gravity.

REEF
A type of reservoir trap composed of rocks, usually limestone, made up of the skeletal remains of marine animals. Reef reservoirs are often characterized by high initial production which falls off rapidly, requiring pressure maintenance techniques to sustain production.

REEL BARGE
A pipe-laying barge equipped with a gigantic reel on which line pipe up to

12 inches in diameter is spooled at a shore station. To lay the pipe, it is unspooled, run through straightening mandrels, inspected, and paid out over the stern of the barge in the manner of a hawser.

REENTRY
To reestablish contact with the well's bore hole in offshore waters, after having moved off location because of weather or other reasons halting drilling operations. A notable example of reentering was that of the Deep Sea Drilling Program by the Scripps Institution of Oceanography when the crew of the drillship Glomar Challenger reentered the hole nine times while drilling in 14,000 feet of water in the Atlantic. See Acoustic reentry.

REEVING A LINE
To string up a tubing or other line in preparation for hoisting; to run a line from the winch up and over a sheave in the crown block and down to the derrick floor.

REFINER-MARKETER
A marketer of gasoline and/or heating oils who operates his own refinery.

REFORMING PROCESSES
The use of heat and catalysts to effect the rearrangement of certain of the hydrocarbon molecules without àltering their composition appreciably; the conversion of low-octane gasoline fractions into higher octane stocks suitable for blending into finished gasoline; also the conversion of naphthas to obtain more volatile product of higher octane number.

RELIEF VALVE
A valve that is set to open when pressure on a liquid or gas line reaches a predetermined level; a pop-off valve.

RENTAL, DELAY
Payment of a sum of money by lessee to the lessor to delay the drilling of a well.

REPEATER STATION
An electronic installation, part of a surveillance and control system for offshore or other remote production operations.

REPRESSURING OPERATION
The injection of fluid into a reservoir whose pressure has been largely depleted by producing wells in the field. This secondary recovery technique used to increase the reservoir pressure in order to recover additional quantities of oil. See Service well.

RE-REFINED OIL
Reclaimed oil (q.v.)

RESERVE PIT
An excavation connected to the working mud pits of a drilling well to hold excess or reserve drilling mud; a standby pit containing already-mixed drilling mud for use in an emergency when extra mud is needed.

RESERVOIR
A porous, permeable sedimentary rock formation containing quantities of oil and/or gas enclosed or surrounded by layers of less permeable or impervious rock; a structural trap; a stratigraphic trap (q.v.).

RESERVOIR PRESSURE
The pressure at the face of the producing formation when the well is shut in. It is equal to the shut-in pressure (at the wellhead) plus the weight in pounds of the column of oil in the hole.

RESID MARKET
The market for residual oils; black oils market.

RESIDUALS
A term used to describe oils that are "leftovers" in various refining processes; heavy black oils used in ships' boilers and in heating plants.

RESIDUE GAS
Gas that remains after processing in a separator or a plant to remove liquids contained in the gas when produced. *See* Tail gas.

RETROGRADE GAS CONDENSATE
A liquid hydrocarbon (condensate) formed in deep formations as the reservoir pressure is reduced through production of natural gas. As the pressure is reduced, the gas condenses to form a liquid instead of the usual pattern of liquid changing to gas. Hence the term "retrograde gas condensate." As liquefaction occurs, the formation rock is "wet" by the condensate which is then not as recoverable as when it was in a gaseous state.

REWORKING A WELL
To restore production where it has fallen off substantially or ceased altogether, cleaning out an accumulation of sand and silt from the bottom of the well.

RHABDOMACY
The "science" of divination by rods, wands, and switches. *See* Doodlebug.

RHEOLOGY
The science that treats of the flow of matter. Rheology in drilling refers to the makeup and handling of a drilling mud circulation system; drilling mud control and characteristics.

RIG
(1) A drilling rig (q.v.). (2) A large tractor-trailer.

RIG BUILDER
(1) A person whose job is to build (or in a modern context) to assemble a derrick. Steel derricks are erected by bolting parts together. (2) Originally, a person who built derricks on the spot out of rig timbers and lumber on which he used cross-cut saws, augurs, axes, hammers, and the adz to fit the wood to his pattern.

RIG MANAGER
One who supervises all aspects of offshore rig operation. Large semi-submersibles, anchored miles at sea with hundreds of workers, are much like a small town engaged in drilling a well in hundreds of feet of water. The rig manager is the resident boss of this floating microcosm.

RIGHT-OF-WAY
(1) A legal right of passage over another person's land. (2) The strip of land for which permission has been granted to build a pipeline and for normal maintenance thereafter. The usual width of right-of-way for common-carrier pipelines (q.v.) is 50 feet.

RIGHT-OF-WAY GANG
A work crew that clears brush, timber, and other obstructions from the right-of-way. The crew also installs access gates in fenced property. *See* Dress-up crew.

RIGHT-OF-WAY MAN
A person who contacts landowners, municipal authorities, government agency representatives for permission to lay a pipeline through their property or through the political subdivision. He also arranges for permits to cross navigable waterways, railroads, and highways from the proper authorities.

RIG REGISTER
A roster of offshore drilling equipment—jackups, semi-submersibles, drill ships, platforms, tenders, and drilling barges—deployed around the world. The register, a modern Jane's Fighting Ships as it were, was introduced by the *Petroleum Engineer* magazine. It is kept current and lists the vessel's or platform's depth capability, equipment, whether self-propelled or towed, and other pertinent information.

RIG TIMBERS
Large-dimension wooden beams used to support the derrick, drilling engines, or other heavy equipment, heavy, roughcut timbers used in the trade by rig builders when derricks were built rather than assembled.

RISER
(1) A pipe through which liquid or gas flows upward. (2) In offshore drilling by semi-submersible, jackup, fixed platform, or drill ship, a riser is the casing extending from the drilling platform through the water to the sea bed through which drilling is done. *See* Marine riser system.

RIVER CLAMPS
Heavy, steel weights made in two halves bolted on screw-pipe at each collar to strengthen the joints and keep the line lying securely on the river bottom or in a dredged trench.

RIVET BUSTER
An air-operated (pneumatic) chisel-like tool for cutting off rivet heads. Used by tankies when tearing down an old tank or other vessel put together with rivets.

ROCK A WELL
To agitate a "dead" well by alternately bleeding and shutting in the pressure on the casing or tubing so that the well will start to flow.

ROCKERS
A counterbalance installed on a shackle-rod line, operating a pumping jack, to pull the rod line back after its power stroke. Rod lines can only pull so on the return stroke the line is kept taught by a counterbalance. Rockers often are in the shape of a box or crate filled with rocks. One edge of the box is attached to a fulcrum-bearing on which it moves back and forth like a rocking chair.

ROCK HOUND
A geologist; a humorous but affectionate colloquialism for a person who assiduously pursues rock specimens in a search for evidence of oil and gas deposits.

ROCK PRESSURE
An early-day term for a well's shut-in or wellhead pressure when all valves are closed and the pressure is observed at the surface.

ROD
(1) Sixteen and one-half feet; the unit of measure used in buying certain types of pipeline right-of-way. (2) A sucker rod; an engine's connecting rod; a piston rod.

RODDAGE FEE
The fee paid to a landowner for the easement of a pipeline right-of-way across his property. Right-of-way is measured in rods (16½ feet) hence the term roddage fee. *See* Right-of-way.

ROD HANGER
A rack with finger-like projections on which rods are hung when pulled from the well; a vertical rack for hanging lengths of pumping rods.

ROD JOB
See Pulling rods

ROD LINE
See Shackle-rod line

ROD-LINE PUMP
See Pump, jerker

ROD PUMPS
See Pumps, rod

ROLL A TANK
To agitate a tank of crude oil with air or gas for the purpose of mixing small quantities of chemical with the oil to break up emulsions or to settle out impurities.

ROLLER BIT
The rock-cutting tool on the bottom of the drill string made with three or four shanks welded together to form a tapered body. Each shank supports a cone-like wheel with case-hardened teeth that rotate on steel bearings.

ROLLING PIPE
Turning a joint of screw-pipe into the coupling of the preceding joint by the use of a rope looped once around the pipe and pulled by a rope crew. This procedure was used on larger diameter line pipe—10 and 12-inch—to make up the connection rapidly before the tongs were put on the pipe for the final tightening.

ROOF ROCK
A layer of impervious rock above a porous and permeable rock formation that contains oil and gas.

ROOT RUN
The first course of metal laid on by a welder in joining two lengths of pipe or other elements of construction; the stringer bead (q.v.).

ROPE SOCKET
A device for securing the end of a steel cable into a connecting piece—a clevis, hook or chain. A metal cup or socket (like a whip socket) into which the cable end is inserted and which then is filled with molten lead or babbitt (q.v.).

ROTARY PUMP

A positive displacement pump consisting of rotary elements—cams, screws, gears, or vanes—enclosed in a case; employed, usually, in handling small volumes of liquid at either high or low pressures. Because of the close tolerances in the meshing of the gears or cams, rotary pumps cannot handle liquids contaminated with grit or abrasive material without suffering excessive wear or outright damage.

ROTARY REAMER

A rock-cutting tool inserted in the drill column just above the drill bit for the purpose of keeping the hole cut to full diameter. Often in drilling deep, hard-rock formations the bit will become worn or distorted, thus cutting less than a full hole. The following reamer trims the hole wall, maintaining full diameter.

ROTARY RIG

A derrick equipped with rotary drilling equipment, i.e., drilling engines, draw works, rotary table, mud pumps, and auxiliary equipment; a modern drilling unit capable of drilling a bore hole with a bit attached to a rotating column of steel pipe.

ROTARY TABLE

A heavy, circular casting mounted on a steel platform just above the derrick floor with an opening in the center through which the drill pipe and casing must pass. The table is rotated by power transmitted from the draw works and the drilling engines. In drilling, the kelly joint fits into the square opening of the table. As the table rotates, the kelly is turned, rotating the drill column and the drill bit.

ROUGHNECKS

Members of the drilling crew; the driller's assistants who work on the derrick floor, up in the derrick racking pipe, tend the drilling engines and mud pumps, and on "trips" operate the pipe tongs breaking out or unscrewing the stands of drill pipe.

ROUND-POINT SHOVEL

A digging tool whose blade is rounded and tapers to a point in the center of the cutting edge. A long-handled shovel, standard equipment for digging ditches by hand.

ROUND TRIP

Pulling the drill pipe from the hole to change the bit and running the drill pipe and new bit back in the hole. On deep wells, round trips or "a trip," as it is more commonly called, may take 24 hours, three 8-hour shifts.

ROUSTABOUT
A production employee who works on a lease or around a drilling rig doing manual labor.

ROYALTY
See Landowner royalty

ROYALTY, COMPENSATORY
Payments to royalty owners as compensation for loss of income which they may suffer due to the failure of the operator to develop a lease properly.

ROYALTY, FIXED-RATE
Royalty calculated on the basis of a fixed rate per unit of production, without regard for the actual proceeds from the sale of the production.

ROYALTY, GUARANTEED
The minimum amount of royalty income a royalty owner is to receive under the lease agreement, regardless of his share of actual proceeds from the sale of the lease's production.

ROYALTY, LANDOWNER'S
A share of the gross production of the oil and gas on a property by the landowner without bearing any of the cost of producing the oil or gas. The usual landowner's royalty is one-eighth of gross production.

ROYALTY OIL
Oil owned by the government, federal, state or local.

ROYALTY, SHUT-IN
Payment to royalty owners under the terms of a mineral lease which allows the operator or lessee to defer production from a well which is shut in for lack of a market or pipeline connection.

RUN
A transfer of crude oil from a stock tank on a production lease to a pipeline gathering system, for transportation to the buyer's facilities; running oil from a tank into a pipeline for delivery to a purchaser.

RUNNING THE TOOLS
Putting the drill pipe, with the bit attached, into the hole in preparation for drilling.

RUN TICKET
A record of the oil run from a lease tank into a connecting pipeline. The ticket is made out in triplicate by the gauger and witnessed by the lease owner's representative, usually the pumper. The run ticket, an invoice for oil delivered, shows opening and closing gauge, API gravity, and temper-

ature, tank temperature, and BS&W. The original of the ticket to the purchaser, a copy to the pumper and one for the gauger.

RUPTURE DISK
A thin, metal plug or membrane in a fitting on a pressure line made so as to blow out or rupture when the pressure exceeds a predetermined level; a safety plug. *See* Soft plug.

S

SADDLE
A clamp, fitted with a gasket, for stopping the flow of oil or gas from holes or splits in a pipeline; a device for making temporary repairs to a line. The clamp conforms to the curve of the pipe and is held in place by U-bolts that fit around the pipe and extend through the clamp.

SADDLE BEARING
A broad, heavy bearing located on top of the Samson post to support the walking beam on a cable tool drilling rig or an oil well pumping jack.

SAE
Society of Automotive Engineers

SAE NUMBER
A classification of lubricating oils in terms of viscosity only. A standard established by the Society of Automotive Engineers. SAE 20; SAE 10W-30, multi-viscosity lubricating oil (q.v.).

SAFETY VALVE
See Relief valve

SALTBED STORAGE
Thick formations or underground layers of salt in which cavities are mined or leached out with super-heated water for the storage of petroleum products, e.g., heating oils, butane, propane, and other LP-gases.

SALT DOME
A subsurface mound or dome of salt. Two types of salt domes are recognized: the piercement and non-piercement. Piercement domes thrust upward into the formations above them causing faulting; non-piercement domes are produced by local thickening of the salt beds and merely lift the overlying formations to form an anticline (q.v.).

SALT-DOME/SALT-PLUG POOLS
Structural or stratigraphic traps (q.v.) associated with rock-salt intrusions; pools formed by the intrusion of underlying salt formations into overlying porous and permeable sedimentary layers creating traps favorable to the presence of oil and gas.

SALT DOME STORAGE
Cavities leached out of underground salt formations by the use of super-heated water for the storage of petroleum products, especially LP-gases.

SALT PLUG
See Salt dome

SAMPLE
Cuttings of a rock formation broken up by the drill bit and brought to the surface by the drilling mud. Rock samples are collected from the shale shaker (q.v.) and examined by the geologist to identify the formation, the type of rock being drilled.

SAMPLE BAG
A small cotton bag with a drawstring to hold rock-cutting samples. Each bag with its sample is tagged with identifying information: well name, lease, location, depth at which cuttings were taken, etc.

SAMPLE LOG
A record of rock cuttings as a well is being drilled, especially in cable-tool drilling. The cuttings, brought to the surface by the bailer, are saved and the depth where obtained is recorded. This record shows the characteristics of the various strata drilled through.

SAMSON POST
A heavy, vertical timber that supports the well's walking beam (q.v.).

SAND
Short for sandstone; one of the more prolific sedimentary rock formations. In informal usage, other sedimentary rocks are also referred to as "sands."

SANDBODY
A sand or sandstone formation defined by upper and lower layers of impervious rock.

SAND CONTROL
A technique for coping with sand from unconsolidated (loose, unpacked) formations that migrate (drift or wash) into down hole pumping equipment or into the bore hole. *See* Gravel packing.

SANDED UP
A well clogged by sand that has drifted and washed into the well bore from the producing formation by the action of the oil.

SAND LINE
A wire line (cable) used on a drilling rig to raise and lower the bailer or sand pump in the well bore. Logging devices and other light-weight equipment are also lowered into the hole on the sand line.

SAND PUMP
A cylinder with a plunger and valve arrangement used for sucking up the pulverized rock, sand, and water from the bottom of the well bore. More effective than a simple bailer. Shell pump; sludger.

SAND REEL
A small hoisting drum on which the sand line is spooled and used to run the bailer or sand pump on a cable tool rig. The sand reel is powered by contact with the band wheel (q.v.).

SANDS
Common terminology for oil-bearing sandstone formations. (Oil is also found in limestone, shales, dolomite, and other porous rock.) In informal or loose usage, other sedimentary rocks are referred to as "sands."

SAND SEPARATOR
A device for removing "drilled solids," pulverized rock and sand, from drilling mud. The sand separator is used in addition to the shale shaker (q.v.) and by removing most of the abrasive material reduces wear on mud pumps and bits.

SATELLITE PLANT
A facility that supports the main processing plant; a plant that derives its feedstock or raw material from the main processing unit.

SATELLITE PLATFORM
Production platform (q.v.)

SATURATION
(1) The extent to which the pore space in a formation contains hydrocarbons or connate water (q.v.). (2) The extent to which gas is dissolved in the liquid hydrocarbons in a formation.

SATURATION PRESSURE
The pressure at which gas begins to be released from solution in oil. *See* Bubble-point pressure.

SBM SYSTEM
Single-buoy Mooring System (q.v.)

SBR

Initials for synthetic butadiene rubber, the main ingredients of which are derived from petroleum. SBR is used in the manufacture of tires, hose, shoes, other severe-duty products.

SCAT (WELDING) RIG

A rack that carries welding generators, gas bottles (CO_2), and spools of welding wire along the pipe being welded. The rig is powered by a small diesel engine. Automatic welding heads ("bugs") are moved ahead on the pipe as the joints are welded.

SCHEDULER

A person in an oil-dispatching office who plans the future movement of batches of crude oil or product in a pipeline system, keeping batches separated and making arrangements for product input and down-stream deliveries. *See* Dispatcher.

SCHLUMBERGER (Slum-ber-jay)

Trade name of a pioneer electrical well-surveying company. In many areas it is common practice to speak of an electric well log as a "slumber-jay" even though the log was made by another company.

SCOURING

The erosion or washing away of the sand/clay covering of a buried subsea pipeline. Scouring caused by sea currents is a serious problem for under-sea lines. Excessive scouring causes spanning, the hanging of a section of the line one to several feet off bottom. If allowed to go uncorrected the pipeline welds crack or the pipe ruptures from its unsupported weight. Subsea lines are inspected for scouring and spanning by side-scan sonar devices or by diver inspection.

SCOUT

A person hired by an operator or a company to seek out information about activities of drilling wells in an area, survey data, drilling rates and depths, and well potentials.

SCOUTING A WELL

Gathering information, by all available means, about a competitor's well—the depth, formations encountered, well logs, drilling rates, leasing, and geophysical reports.

SCRAPER, PIPELINE

A pig; a cylindrical, plug-like device equipped with scraper blades, wire brushes, and toothed rollers used to clean accumulations of wax, rust, and other foreign matter from pipelines. The scraper is inserted in the line at a "trap" (q.v.) and is pushed along by the pressure of the moving column of oil.

SCRAPER TRAP
A facility on a pipeline for inserting and retrieving a scraper or "pig." The trap is essentially a "breech-loading" tube isolated from the pipeline by valves. The scraper is loaded into the tube like a shell into a shotgun; a hinged plug is closed behind it, and line pressure is then admitted to the tube *behind* the scraper. A valve is opened ahead of the scraper and it is literally pushed into the line and moved along by the oil pressure.

SCREW CONVEYOR
A mechanism for moving dry, solid material—pelletized plastics, sulfur, cement, etc.—from one location to another by means of a helix or screw rotating in a cylindrical conduit. Archimedes thought of it first.

SCRUBBING
Purifying a gas by putting it through a water or chemical wash; also the removal of entrained water.

SCRUBBING PLANT
A facility for purifying or treating natural gas for the removal of hydrogen sulfide or other impurities.

SEA TERMINAL
An offshore loading or unloading facility for large, deep-draft tankers. The terminal is served by filling lines from shore or by smaller, shallow-draft vessels.

SEALINES
Submarine pipelines; lines laid on the ocean floor from offshore wells to production platform, and to receiving stations on shore.

SEALS
Thin strips of metal, imprinted with serial numbers, used to "seal" a valve in an open or closed position. The metal strip has a locking-snap on one end into which the free end is inserted, locking it securely. Seals are used on tanks in a battery to prevent the undetected opening or closing of a valve.

SEAMLESS PIPE
Pipe made without an axial seam; pipe made from a billet or solid cylinder of hot steel and "hot-worked" on a mandrel into a tubular piece without a seam. *See* Lap welded.

SECONDS SAYBOLT FUROL (SSF)
A measurement of the viscosity of a heavy oil. Sixty cubic centimeters of an oil are put in an instrument known as a "Saybolt viscosimeter" and permitted to flow through a standardized orifice in the bottom at a specified temperature. The seconds required to flow through is the oil's viscosity, its SSF number. *See* also Seconds saybolt universal.

SECONDS SAYBOLT UNIVERSAL
A measurement of the viscosity of a light oil. A measured quantity of oil—usually 60 cubic centimeters—is put in an instrument known as a "Saybolt viscosimeter" and permitted to flow through an orofice in the bottom at a specified temperature. The number of seconds required for the flow-through is the oil's SSU number, its viscosity.

SECTION OF LAND
One square mile; 640 acres; sixteen 40-acre plots.

SECONDARY RECOVERY
The extraction of oil from a field beyond what can be recovered by normal methods of flowing or pumping; the use of water flooding, gas injection, and other methods to recover additional amounts of oil.

SEDIMENTARY BASIN
An extensive area (often covering thousands of square miles) where substantial amounts of unmetamorphized sediments occur. Most sedimentary basins are geologically depressed areas (shaped like a basin). The sediment is thickest in the interior and tends to thin out at the edges. There are many kinds of such basins, but it is in these formations that all the oil produced throughout the world has been found.

SEDIMENTARY ROCK
Rock formed by the laying down of matter by seas, streams, or lakes; sediment (mineral fragments, animal matter) deposited in bodies of water through geologic ages. Limestone, sandstone, shale are sedimentary rocks.

SEISMIC SEA-STREAMER
A cable, trailed from a geophysical vessel, towing a series of hydrophones along the sea floor recording seismic "signals" from underwater detonations. As the vessel moves slowly ahead, harmless electronic or air detonations are set off which are reflected from rock formations beneath the sea floor and picked up by the sensitive, sound-detecting hydrophones. *See* Geophone.

S.E.G.
The Society of Exploration Geophysicists, a professional organization of geophysicists engaged in exploration for oil and gas.

SEISMIC "THUMPER"
See Vibrator vehicle

SEISMOGRAPH
A device that records vibrations from the earth. As used in the exploration for oil and gas, a seismograph records shock waves set off by explosions detonated in shot holes (q.v.) and picked up by geophones (q.v.).

SEISMOGRAPHIC SURVEY
Geophysical information on subsurface rock formations gathered by means of a seismograph (q.v.); the investigation of underground strata by recording and analyzing shock waves artificially produced and reflected from subsurface bodies of rock.

SEISMOMETER
A device for receiving and recording shock waves set off by an explosion or other seismic sources and reflected by underground rock formations.

SEMI-DIESEL
A misnomer for a diesel-cycle engine whose compression is not high enough to create sufficient heat to ignite the injected fuel when starting cold. Semi-diesels or, more correctly, hot-head or hot-plug diesels, are equipped with a plug that extends into the firing chamber heated by a torch or by electricity to assist in the ignition of the diesel fuel until the engine is running and up to operating temperature; a small, low-compression diesel engine. *See* Hot plug engine.

SEPARATOR
A pressure vessel (either horizontal or vertical) used for the purpose of separating well fluids into gaseous and liquid components. Separators segregate oil, gas, and water with the aid, at times, of chemical treatment and the application of heat.

SEPARATOR GAS
Natural gas separated out of the oil by a separator at the well.

SERVICE TOOLS
A variety of down hole equipment used in drilling, completion, and work-over of oil and gas wells.

SERVICE WELL
A non-producing well used for injecting water or gas into the reservoir or producing formation in pressure maintenance or secondary recovery programs; also a salt-water disposal well.

SERVO
Short for servomechanism (q.v.).

SERVOMECHANISM
An automatic device for controlling large amounts of power with a small amount of force. An example of a servomechanism is the power-steering on an automobile. A small force on the steering wheel activates a hydraulically-powered mechanism that does the real work of turning the wheels.

SERVOMOTOR
A power-driven mechanism that supplements a primary control operated by a comparatively small force. *See* Servomechanism.

SET BACK
The space on the derrick floor where stands of drill pipe or tubing are "set back" and racked in the derrick. Off-shore drilling platforms often list the stand capacity of their set backs as an indication of their pipe-handling capability and capacity. On transportable, mast-type derricks used on land, set backs are outside the derrick proper.

SET CASING, TO
To cement casing in the hole. The cement is pumped down hole to the bottom of the well and forced up a certain distance into the annular space between casing and the rock wall of the drill hole. It is then allowed to harden, thus sealing off upper formations that may contain water. The small amount of cement in the casing is drilled out in preparation for perforating (q.v.) to permit the oil to enter the casing. The decision to set casing (or pipe) is an indication that the operator believes he has a commercial well.

SETTLED PRODUCTION
The lower average production rate of a well after the initial flush production (q.v.) tapers off; the production of a well that has ceased flowing and has been put on the pump.

SEVERANCE TAX
A tax levied by some states on each barrel of oil or each thousand cubic feet of gas produced. Production tax.

SHACKLE ROD
Jointed steel rods, approximately 25 feet long and ¾ to 1 inch in diameter used to connect a central power (q.v.) with a well's pumping unit or pumping jack. Shackle-rod lines are supported on metal posts (usually made of 2-inch line pipe) topped with wooden guide blocks which are lubricated with a heavy grease.

SHAKE OUT
To force the sediment in a sample of oil to the bottom of a test tube by whirling the sample at high speed in a centrifuge machine. After the sample has been whirled for three to five minutes, the percent of BS&W (sediment and water) is read on the graduated test tube.

SHALE
A type of sediment rock composed of fine particles of older rock laid down as deposits in the waters of lakes and seas. Most shales are compacted muds and consequently do not contain oil or gas in commercial quantities.

SHALE OIL
Oil obtained by treating the hydrocarbon kerogen found in certain kinds of shale deposits. When the shale is heated the resulting vapors are condensed and then treated in an involved process to form what is called shale oil or synthetic oil.

SHALE SHAKER
A vibrating screen for sifting out rock cuttings from drilling mud. Drilling mud returning from down hole carrying rock chips in suspension flows over and through the mesh of the shale shaker leaving the small fragments of rocks which are collected and examined by the geologist for information on the formation being drilled.

SHAPED-CHARGE PERFORATION
A perforation technique using shaped explosive charges instead of steel projectiles to make holes in casing. Quantities of explosives are made in special configurations and detonated at the bottom of the hole against the casing wall to make the perforations.

SHARPSHOOTER
A spade; a narrow, square-ended shovel used in digging. Sharpshooters are one of the pipeliner's digging tools used for squaring up a ditch or the sides of a bell hole (q.v.).

SHAVE-TAILS
A skinner's (q.v.) term for his mules.

SHEAVE
A grooved pulley or wheel; part of a pulley block; a sheave can be on a fixed shaft or axle (as in a well's crown block) (q.v.) or in a free block (as in block and tackle).

SHEET IRON
Galvanized, corrugated sheet metal used for roofing, garages, and other more or less temporary buildings. Because a sheet iron or corrugated iron building is relatively inexpensive and easy to assemble, this kind of construction is common on oil leases.

SHELL PUMP
See Sand pump

SHIMS
Thin sheets of metal used to adjust the fit of a bearing or to level a unit of equipment on its foundation. For fitting a bearing, a number of very thin (.001 to .030-inch) shims are put between the two halves of the bearing (between the box and cap). Shims are added or removed until the bearing fits properly on the journal.

SHIRT TAILS
Colloquial term for the structural members or shanks of a drill bit that anchor the cutting wheels; the frame of the bit below the threaded pin.

SHOESTRING SAND
Narrow strands of saturated formation that pinch out (q.v.) or are bounded by less permeable strata that contain no oil.

SHOOT A WELL, TO
To detonate an explosive charge in the bottom of a well to fracture a "tight" formation in an effort to increase the flow of oil. *See* Well shooter.

SHOOTER
See Well shooter

SHOOTING LEASE
An agreement granting permission to conduct a seismic or geophysical survey. The lease may or may not give the right to lease the land for oil or gas exploration.

SHORT TRIP
Pulling the drill string part way out of the hole. Short trips may be necessary to raise the drill up into the protective string of casing to avoid having the drill string stuck in the hole by a cave-in or sloughing of the wall of the bore hole below the protective casing.

SHOT-GUN TANK
A tall, slender tank for separating water and sediment from crude oil. *See* Boot.

SHOT HOLE
A small-diameter hole, usually drilled with a portable, truck-mounted drill, for "planting" explosive charges in seismic operations.

SHOT POINT
The shot hole (q.v.); the point at which a detonation is to be made in a geophysical survey.

SHOW OF OIL
A small amount of oil in a well or a rock sample; a show of oil usually signifies the well will not be a commercial producer.

SHUT-DOWN—SHUT-IN WELL
There is a great difference between a shut-down and a shut-in. A well is shut down when drilling ceases which can happen for many reasons: failure of equipment; waiting on pipe; waiting on cement; waiting on orders from the operator, etc. A well is shut in when its wellhead valves are closed, shutting off production. A shut-in well often will be waiting on tankage or a pipeline connection.

SHRINK FIT
An extremely tight fit as the result of "shrinking" one metal part around another. A heated part is placed around a companion piece, and as the heated part cools, a shrink fit results. Conversely, an expansion fit may be made by cooling a part (a valve-seat insert, for example) to extremely low temperature with "dry ice" and placing the part in position. As it returns to normal temperature, a tight "expansion fit" will result.

SHUT IN
To close the valves at the wellhead so that the well stops flowing or producing; also describes a well on which the valves have been closed.

SHUT-IN PRESSURE
Pressure as recorded at the wellhead when the valves are closed and the well is shut in.

SHUT-IN ROYALTY
Payments made when a gas well, capable of producing in paying quantities, is shut in for lack of market for the gas. This type of royalty or some form of rental is usually required to prevent termination of the lease.

SIDE-BOOM CATS
See Boom-cats

SIDE-DOOR ELEVATORS
Casing or tubing elevators (q.v.) with a hinged latch that opens on one side to permit it to be fastened around the pipe and secured for hoisting.

SIDE TRACKING
Drilling of another well beside a non-producing well and using the upper part of the non-producer. A method of drilling past obstructions in a well, i.e., lost tools, pipe, or other material blocking the hole.

SIDE-WALL CORE/SAMPLE
A sample of rock taken from the wall of the well's bore hole.

SIGHT GLASS
A glass tube in which the height of a liquid in a tank or pressure vessel may be observed. The glass tube is supported by fittings that extend through the vessel wall thus allowing the fluid in the tank to assume a corresponding level in the glass.

SIGHT PUMP
An "antique" gasoline dispensing system in which the gasoline was pumped by hand into a ten-gallon glass tank atop the pump in plain sight of the customer. When the glass cylinder had been pumped full, the attendant opened the valve on the filling hose which permitted the gasoline to gravitate into the vehicle's tank. Gravity pump.

SIGMA
Society of Independent Gasoline Marketers of America

SILENCER
A large, cylindrical vessel constructed with an arrangement of baffles, ports, and acoustical grids to muffle the exhaust noises of stationary engines.

SINGLE-BUOY MOORING SYSTEM
An offshore floating platform (20 to 35 feet in diameter) connected to pipelines from the shore for loading or unloading tankers. The SBM system is anchored in deep water thus permitting large tankers to offload or "lift" cargo in areas where it is impractical to build a loading jetty or the close-in water is too shallow for deep-draft vessels.

SINGLE-POINT MOORING
Single-buoy mooring (q.v.)

SITTING ON A WELL
The vigil of the geologist, the operator, and other interested parties who literally sit waiting for the well's drill to bore into what is expected to be the producing formation. The geologist examines the cuttings brought up by the drilling mud to ascertain just when the pay zone is penetrated. On a "big well," a very good well, everyone knows when the pay is reached; on small or marginal wells, the geologist may be the only one who recognizes it.

SIZEING SCRAPER
A cylindrical plug-like tool that is pushed or pulled through a length of pipe to test for roundness.

SKID-MOUNTED
Refers to a pumping unit or other oil field equipment that has no permanent or fixed foundation but is welded or bolted to metal runners or timber skids. Skid-mounted units are usually readily movable by pulling as a sled or by hoisting onto a truck.

SKIDDING THE RIG
Moving the derrick from one location to another without dismantling the structure; transporting the rig from a completed well to another location nearby by the use of skids (heavy timbers), rollers, and a truck or tractor. Transportable folding or jack-knife rigs are seldom skidded; they are folded down to a horizontal position and moved on a large, flatbed truck.

SKIDS
Squared, wooden timbers used to support line pipe while it is being welded; any rough-cut lumber used to move or support a heavy object.

SKID TANK
A product-dispensing tank mounted on skids or runners. Can be pulled or carried on a truck.

SKIMMER
A type of oil spill clean-up device propelled over the water that sucks or paddles the oil into a tank.

SKIMMING PLANT
(1) A topping plant (q.v.); (2) A facility built alongside a creek or small stream to catch and skim off oil that, in the early days in some fields, was turned into creeks or accidentally discharged from lease tanks or from broken pipelines.

SKINNER
See Mule skinner

SLAB PATCH
A metal patch made out of a section of pipe welded over a pitted or corroded section of pipeline. *See* Half-sole.

SLANT-HOLE TECHNIQUE
A procedure for drilling at an angle from the vertical by means of special down-hole drilling tools to guide the drill assembly in the desired direction. Slant holes are drilled to reach a formation or reservoir under land that can not be drilled on, such as town site, beneath a water-supply lake, a cemetery or industrial property where direct, on-site drilling would be impractical or unsafe. Slant holes also are drilled to flood a formation with water or mud to kill a wild or burning well. *See* Killer well.

SLANT WELL
Directional well (q.v.).

SLEEVE FITTING
A collar or nipple that is slipped over a length of pipe to repair a leak caused by a split or corrosion. When the sleeve is in place, the ends are welded to the pipe beyond the damaged section.

SLIDE VALVES
Very large, box-like valves for flues and stacks. Made from sheet steel, the valves are mechanically or hydraulically operated.

SLIM-HOLE DRILLING
A means of reducing the cost of a well by drilling a smaller diameter hole than is customary for the depth and the types of formations to be drilled through. A slim hole permits the scaling down of all phases of the drilling and completion operations, i.e., smaller bits, less powerful and smaller rigs (engines pumps, draw works) smaller pipe and less drilling mud.

SLIP

A horse-drawn, earth-moving scoop. The slip has two handles by which the teamster guides the metal scoop into the ground at a slight angle to skim off a load of earth. Teams and slips were used to dig slush pits and build tank dikes before the days of the bulldozer. A full slip would hold about one-half cubic yard.

SLIP JOINT

A special sleeve-like section of pipe run in the drill string to absorb the vertical motion of a floating drilling platform caused by wave action; a heave compensator (q.v.).

SLIP LOAD

The weight of the string of drill pipe, tubing or casing suspended in the drill hole by the slips (q.v.).

SLIPS

Wedge-shaped pieces of metal that fit into a bushing in the rotary table to support the string of tubing or drill pipe.

SLIP STICK

An engineer's slide rule; a log-log rule; an instrument consisting of a ruler and a medial slide graduated with logarithmic scales used for rapid calculations.

SLOP TANK

(1) On a products pipeline, a tank where off-specification products or interface-mix is stored. (2) At a marine terminal, a tank for holding the oil/water mix from a vessel that has washed down its compartments. (3) Any vessel used for retaining contaminated oil or water until it can be properly disposed of.

SLUDGE

An oleo-like substance caused by the oxidation of oil or by contamination with other material; a thick, heavy emulsion containing water, carbon, grit, and oxidized oil.

SLUDGER

See Sand pump

SLUSH-PIT LAUNDER

A wooden or metal square-sided conduit or sluice box where the bailer is dumped, the water, mud, and rock chips flushing down the launder into the slush pit. This device, a cousin to the launder used in washing ore from a mine, is part of a cable-tool drilling scene.

SLUSH PUMP

Mud pump (q.v.)

SMOKELESS FLARE
A specially constructed vertical pipe or stack for the safe disposal of hydrocarbon vapors or, in an emergency, process feed that must be disposed of. Smokeless flares are equipped with steam jets at the mouth of the stack to promote the complete combustion of the vented gases. The jets of steam induce greater air flow and cools the flame resulting in complete combustion without smoke or ash.

SNAP GRABBER
A member of a work gang who manages to find easy jobs to keep himself busy while the heavy work is being done by his companions. A fully occupied loafer.

SNATCH BLOCK
A block whose frame can be unlatched to insert a rope or wire line; a single-sheave block used more often for horizontal pulling than for hoisting with "A" frame or mast.

SNG
Synthetic natural gas; gas manufactured by various processes from coal, tar sands, or kerogen shales (q.v.). Substitute natural gas.

SNOW-BANK DIGGING
Colloquial expression for the relatively soft, easy drilling in sand, shales, or gumbo.

SNUBBING
A procedure for servicing wells that are under pressure. Tubing, packers, and other down hole tools are withdrawn from the well through a stack of rams (valve-like devices that close around pipe or tubing being withdrawn and seal off the well pressure). As each joint of tubing is withdrawn, it is unscrewed.

SOFT PLUG
A safety plug in a steam boiler, soft enough to give way or blow before the boiler does from excessive high pressure; the plug in an engine block that will be pushed out in case the cooling water in the block should freeze, thus preventing the ice from cracking the block.

SOFTWARE
The collection of programs used in a particular application for use in a computer. Tapes, cards, disk packs containing programs designed for a process or series of processes.

SOLENOID
An electrical unit consisting of a coil of wire in the shape of a hollow cylinder and a moveable core. When energized by an electric current, the coil acts as a bar magnet, instantly drawing in the moveable core. A

solenoid on an automobile's starting mechanism causes the starter-motor gear to engage the toothed ring on the vehicle's flywheel, turning the engine. Solenoids are used also for opening and closing quick-acting, plunger-type valves, as those on washing machines and automatic dishwashers.

SOLUTION GAS
Natural gas dissolved and held under pressure in crude oil in a reservoir. *See* Solution-gas field.

SOLUTION-GAS FIELD
An oil reservoir deriving its energy for production from the expansion of the natural gas in solution in the oil. As wells are drilled into the reservoir, the gas in solution drives the oil into the well bore and up to the surface.

SOLVENT
A liquid capable of absorbing another liquid, gas, or solid to form a homogenous mixture; a liquid used to dilute or thin a solution.

SONIC INTERFACE-DETECTOR
A pipeline sensing "probe" for detecting the approach of a product interface by identifying the change in sound velocities between the two products being pumped. The electronic device has a probe inserted through the wall of the pipeline, protruding into the fluid stream. The probe picks up the variations in sound velocities, and through the proper linkage can given an audible alarm or actuate valves when the interface arrives.

SOUP
Nitroglycerine used in "shooting" a well. Nitro in its pure form is a heavy, colorless, oily liquid made by treating glycerin with a mixture of nitric and sulfuric acids. It is usually mixed with absorbents for easier handling. Nitro, when used in well shooting, is put in tin "torpedos," 4 to 6 inches in diameter, and lowered into the well on a line. The bottom of each torpedo can is made to nest in the top of the preceding one, so as many cans as necessary for the shot can be lowered in and stacked up. Nitro is measured in quarts; the size of the shot depends upon the thickness and hardness of the formation to fractured.

SOUR GAS
Natural gas containing chemical impurities, notably hydrogen sulfide (H_2S) or other sulfur compounds that make it extremely harmful to breathe even small amounts; a gas with a disagreeable odor resembling that of rotten eggs.

SOUR PRODUCTS
Gasoline, naphthas, and refined oils which contain hydrogen sulfide (H_2S) or other sulfur compounds. Sourness is directly connected with odor.

SOUR-SERVICE TRIM
A designation by manufacturers of oil field fittings and equipment that their products have finishes resistant to corrosion by hydrogen sulfide (H_2S) and other corrosive agents in "sour" oil and gas. See Sour gas.

SPACING PATTERN
Geographic subdivision established by governmental authority, usually state, defining the number of acres to be allotted to each well drilled in a common reservoir.

SPEARS
Fishing tools for retrieving pipe or cable lost in the bore hole. Some spears resemble harpoons with fixed spurs, others have retractable or releasing-type spurs.

SPHEROID
As it applies to the industry, a spheroid is a steel storage tank in the shape of a sphere flattened at both "poles," designed to store petroleum products, mainly LP-gases, under pressure. See Hortonsphere.

SPIDERS
The hinged, latching device attached to the elevators (the hoisting arms that lift pipe and casing in the derrick). Elevators-spider is a unit, and is attached to the travelling-block hook for hoisting pipe, casing, and tubing out of the hole and lowering in. The spider is manually locked around a length of tubing just below the tool joint. Some advanced types of elevator spiders are air operated.

SPINDLETOP
The name of the gusher brought in by Capt. Anthony Lucas, near Beaumont, Texas in 1901. The well, the first important producer ever drilled with rotary tools, blew in (literally) and produced at the rate of 75,000 to 100,000 barrels a day.

SPINNING CHAIN
A light chain used by the drilling crew on the derrick floor when running and pulling tubing or drill pipe. After a joint has been "broken" or loosened by the pipe tongs, the spinning chain is given several turns around the pipe and when the chain is pulled, the pipe is rotated counter-clockwise and quickly unscrewed.

SPIN UP
To screw one stand of drill pipe or tubing rapidly into another with a spinning chain (q.v.). After making up the joint in this manner, the heavy pipe tongs are applied to make the joint tight.

SPINNING WRENCH
An air-operated drill pipe or tubing wrench used in place of the spinning chain and the winch-operated wrenches.

SPLASH ZONE
The area where waves of ocean or lake strike the support members of offshore platforms and production installations; the water line. The splash zone is particularly subject to corrosion because of the action of both salt water and air.

SPLIT SLEEVE
A type of pipeline repair clamp made in two halves that bolt together to form a pressure-tight seal over a hole or split in the pipe. Split sleeves also are made to enclose leaking valves and flanges until they can be permanently repaired.

SPM
Strokes per minute; indicates the speed or pumping rate of reciprocating pumps.

SPONSON
An air chamber along the sides of a barge or small ship to increase buoyancy and stability. Sponsons are used on crane-barges for additional buoyancy and to minimize listing when heavy, off-side lifts are being made with the crane.

SPOT CHARTER TANKER RATES
The cost per ton to move crude oil or product by tanker from one port to another on a one-time basis, as compared to long-term charter rates. Spot charter rates fluctuate widely with demand and availability of tonnage.

SPREAD
A contractor's men and equipment assembled to do a major construction job, a "spread" may be literal, as the men and equipment are strung out along the right-of-way for several miles. On well work-over, or other jobs, the spread is a concentration of the equipment for the work.

SPUD
To start the actual drilling of a well.

SPUDDER
The name for a small, transportable cable tool drilling rig. Spudders are used in shallow-well workovers, for spudding in, or bringing in a rotary-drilled well.

SQUEEZE A WELL
A technique to seal off with cement a section of the well bore where a leak or incursion of water or gas occurs; forcing cement to the bottom of the

casing and up the annular space between the casing and the wall of the bore hole to seal off a formation or plug a leak in the casing; a squeeze job.

SS-2000 CLASS RIG
The designation for the class of semisubmersible drilling platforms (the largest built to date: 1976) which are of 18,000-ton displacement; 2,000-ton deck-load capacity; and capable of drilling in 2,000 feet of water.

SSU & SSF
Seconds Saybolt Universal and Seconds Saybolt Furol (q.v.).

STABBER
(1) A pipeline worker who holds one end of a joint of pipe and aligns it so that it may be screwed into the collar of the preceding joint. Before the days of the welded line, the pipeline stabber worked only half a day because of the exhausting nature of his work. (2) On a pipe-welding crew, the stabber works the line-up clamps or line-up mandrel. (3) On a drilling rig, the floorman (roughneck) who centers the joint of pipe being lowered into the tool joint (q.v.).

STABBING BOARD
A platform 20 to 40 feet up in the derrick used in running casing. The derrick man stands on the stabbing board and assists in guiding the threaded end of the casing into the collar of the preceding joint that is hanging in the slips in the rotary table.

STABILIZER SLEEVE
A bushing the size of the bore hole inserted in the drill column to help maintain a vertical hole, to hold the bit on course. The bushing or sleeve can be the fixed or rotating type with permanent or replaceable wings or lugs. (The lugs protrude from the body of the sleeve, making contact with the wall of the hole.)

STACK THE TOOLS
Pulling the drill pipe and laying it down (stacking outside the derrick) in preparation for skidding or dismantling the derrick. If the rig is transportable, it is folded down and made ready to move.

STALKS
Colloquialism for joints of line pipe, tubing, or drill pipe.

STANDARD TOOLS
See Cable tools

STAND OF PIPE
A section of drill pipe or tubing (one, two, or three—sometimes, four joints) unscrewed from the string as a unit and racked in the derrick. The height

of the derrick determines the number of joints that can be unscrewed in one "stand of pipe." *See* Doubles.

STANDPIPE
The pipe that conveys the drilling mud from the mud pump to the swivel (q.v.). The standpipe extends part way up the derrick and connects to the mud hose which is connected to the gooseneck (a curved pipe) of the swivel.

STATIC LINE
A wire or line to drain off or ground static electricity that may have built up from friction in a vehicle or its contents; a grounding line for gasoline transports to prevent arcing of static charges when unloading.

STEAMING PLANT
See Treating plant

STEAM PUMP
A reciprocating pump that receives its power from high-pressure steam. Steam is piped into the pump's steam chest and from there it is admitted to the power cylinder where it acts upon the pump's power pistons, driving them to and fro as the steam valves open and close. The fluid end of the pump is driven by the steam pistons. *See* Pump, Simplex.

STEAM TRAP
A device on a steam line designed to trap air and water condensate and automatically bleed the air and drain the water from the system with a minimum loss of steam pressure.

STEEL REEF
Refers to the artificial "reefs" formed by the substructures of offshore drilling and production platforms that attract a variety of marine life from barnacles and algae to many kinds of fish.

STEEL STORAGE
Refers to the storage of crude oil and products in above-ground steel tanks.

STEP-OUT WELL
A well drilled adjacent to a proven well but located in an unproven area; a well located a "step out" from proven territory in an effort to determine the boundaries of a producing formation.

STICK-ELECTRODE WELDING
Electric-arc welding in which the welding rod or electrode is hand-held as compared to automatic welding.

STILE
Steps made for walking up and over a fence or other obstruction. Made in

the shape of the letter A, stiles are used on farms and fenced leases to get to the other side without going through a gate. *See* Cattle guard.

STILL, PIPE
A type of distillation unit in which oil to be heated passes through pipes or tubes in the form of a flat coil, similar to certain kinds of heat exchangers. There are two main chambers in a pipe still, one where the oil is preheated by flue gases (the convection chamber), the other, the radiant-heat chamber raises the oil to the required temperature. No distillation or fractionation takes place in the still proper. The hot oil is piped to a bubble tower or fractionation tower where the oil flashes or vaporizes. The vapors are then condensed into a liquid product.

STILL, SHELL
The oldest and simplest form of a distillation still; a closed vessel in which crude oil is heated and the resulting vapors conducted away to be condensed into a liquid product.

STIMULATION
The technique of getting more production from a downhole formation. Stimulation may involve acidizing, hydraulic fracturing, shooting or simply cleaning out to get rid of and controlling sand.

STINGER
The pipe guide at the laying-end of a lay barge (q.v.). On a reel-type lay barge where the coiled pipe is straightened before being laid over the end of the barge, the stinger controls the conformation of the pipe as it leaves the barge.

STOCK AND DIES
A device for making threads on the end of a joint of pipe or length of rod; an adjustable frame holding a set of steel dies or cutting teeth that is clamped over the end of the pipe to be threaded. When properly aligned the dies are rotated clockwise in the frame, cutting away excess metal, leaving a course of threads.

STOCK TANK
(1) A lease tank into which a well's production is run. (2) Colloquial term for a cattle pond, particularly in the Southwest.

STOP
A common term for a type of plug valve used on lease tanks and low-pressure gravity systems.

STOP-AND-WASTE VALVE
A type of plug valve that when in a closed position drains the piping above or beyond it. When the valve is turned a quarter turn to shut it off, a small port or hole in the valve body is uncovered, permitting water above the

valve to drain out, preventing a freeze up in cold weather. Stop-and-waste valves are used mainly on small-diameter water piping.

STOPCOCK
A type of plug valve usually installed on small-diameter piping; pet cock.

STOPCOCKING
Shutting in wells periodically to permit a buildup of gas pressure in the formations and then opening the wells for production at intervals.

STOPPEL
A plug inserted in a pipeline to stop the flow of oil while repairs are being made; a specially designed plug inserted in a pipeline through the use of a tapping machine (q.v.).

STORAGE JUG
The name applied to underground salt cavities for the storage of LP-gases and other petroleum products. Jug-shaped cavities are leached or washed out of salt beds using super-heated water under pressure. The resulting underground caverns, some are 100 feet in diameter and 900 feet deep, are ideal storage wells for petroleum products. *See* Salt-bed storage.

STORE LEASE
A preprinted lease form (bought at the store) with blanks to be filled in by the parties to the lease.

STORM CHOKE
A safety valve installed in the well's tubing below the surface to shut the well in when the flow of oil reaches a predetermined rate. Primarily used on offshore, bay, or townsite locations, the tubing valve acts as an automatic shut-off in the event there is damage to the control valve or the Christmas tree.

STOVE OIL
A light fuel oil or kerosine used in certain kinds of wickless-burner stoves.

STRADDLE PLANT
See On-line plant

STRAIGHT-RUN
Refers to a petroleum product produced by the primary distillation of crude oil; the simple vaporization and condensation of a petroleum fraction, without the use of pressure or catalysts.

STRAIN GAUGE
Any of various devices that measure the deformation of a structural element, pipe, or cable, subject to loads. *See* Pressure tranducer.

STRAPPING
Measuring a tank with the use of a steel tape to arrive at its volume; strapping involves measuring the circumference at intervals, top to bottom; height, steel thickness, and computing deadwood (q.v.). Tank tables (q.v.) are made from these measurements.

STRATIGRAPHIC TEST HOLE
A hole drilled to gather information about a stratigraphic formation, the general character of the rocks, their porosity, and permeability.

STRATIGRAPHIC TRAP
A type of reservoir (q.v.) capable of holding oil or gas, formed by a change in the characteristics of the formation—loss of porosity and permeability, or a break in its continuity—which forms the trap or reservoir.

STRATIGRAPHY
Geology that deals with the origin, composition, distribution, and succession of rock strata.

STRAW IN THE CIDER BARREL
To have a well in a producing reservoir; or to have an interest in a well in a producing field.

STREAM DAY
An operating day on a process unit as opposed to a calendar day. Stream day includes an allowance for regular downtime.

STRIKE
(1) The angle of inclination from the horizontal of an exposed strata of rock.
(2) A good well; to make a strike is to find oil in commercial quantities; a hit.

STRINGER BEAD
A welding term that refers to the first bead or course of molten metal put on by the welder as two joints of line pipe are welded together. See Pipeline welding.

STRINGING PIPE
Placing joints of line pipe end to end along a pipeline right-of-way in preparation for laying, i.e., screwing or welding the joints together to form a pipeline.

STRIP CHART
In lieu of the circular chart for recording gas flow through an orifice meter, strip charts are sometimes used. Strip charts, as long as 35 to 40 feet, need not be changed more than once a month if the operator desires. Also, the speed at which the long chart moves through the meter is adjustable so the recording of fluctuations in gas flow may be spread out, permitting more accurate readings.

STRIPPER
An oil well in the final stages of production; a well producing less than 10 barrels a day. Most stripper wells are pumped only a few hours a day. In 1976 there were nearly 400,000 stripper wells in the U.S. producing 20 percent of the country's oil.

STRIPPER TOWER, SOUR-WATER
A refinery vessel, a tower for the physical removal of contaminants from "sour water," water from knockout drums, condensates from accumulators, and other processing units, before it undergoes biological treatment or is discharged in the plant's waste-water system.

STRIPPING JOB
See Pulling rods

STRIPPING PLANT
See Gasoline plant

STRIPPING THE PIPE
The job of removing drill pipe or tubing from a well under pressure, while maintaining control of the well. The pipe is "stripped" by withdrawing it, a "stand" at a time, through a wellhead plug equipped with a hydraulic closure mechanism (ram) that maintains pressure-contact with the pipe being withdrawn.

STRIPPING THE WELL
To pull the rods and tubing from the well at the same time. The tubing must be "stripped" over the rods, a joint at a time.

STRIP, TO
To disassemble; to dismantle for the purpose of inspection and repair; to remove liquid components from a gas stream. See also Stripping the pipe.

STRUCTURAL TRAP
A type of reservoir containing oil and/or gas formed by movements of the earth's crust which seal off the oil and gas accumulation in the reservoir forming a trap. Anticlines, salt domes, and faulting of different kinds form structural traps. See Stratigraphic trap.

STRUCTURE
Subsurface folds or fractures of rock layers which may form a reservoir capable of holding oil or gas.

STRUCTURE CONTOUR MAP
See Contour map

STUB IN, TO
To attach a line (usually of smaller diameter) to an existing line, manifold,

or vessel and making the connection by cutting a hole in the existing installation and welding on a nipple or other fitting.

STUB LINE
An auxiliary line attached to an existing line by use of a tap saddle (q.v.) or by welding on a nipple or other fitting.

STUD DUCK
Top man; the big boss.

STUFFING BOX
A packing gland; a chamber or "box" to hold packing material compressed around a moving pump rod or valve stem by a "follower" to prevent the escape of gas or liquid.

SUB
A short length of tubing containing a special tool to be used down hole; a section of steel pipe used to connect parts of the drill column which, because of difference in thread design, size or other reason, cannot be screwed together; an adapter.

SUBMERSIBLE BARGE PLATFORM
A type of drilling rig mounted on a barge-like vessel used in shallow coastal waters. When on location, the vessel's hull is submerged by flooding its compartments leaving the derrick and its equipment well above the water line.

SUBMERSIBLE DRILLING-BARGE
A barge-like vessel capable of drilling in deeper water than the smaller and simpler barge platform. The submersible drilling-barge has a drilling deck separate from the barge element proper. When floated into position offshore in water as deep as 100 feet, the barge hull is flooded and as it slowly sinks, the drilling platform is simultaneously raised on jacking-legs at each corner of the barge, keeping the drill platform well above the water surface.

SUB-SEA COMPLETION SYSTEM
A self-contained unit resembling a bathysphere to carry men to the ocean bottom to install, repair, or adjust wellhead connections. One type of modular unit is lowered from a tender and fastened to a special steel, wellhead cellar. Men work in a dry, normal atmosphere. The under-water wellhead system was developed by Lockheed Petroleum Services Ltd. in cooperation with Shell Oil Company.

SUBSTRUCTURE
The sturdy platform upon which the derrick is erected. Substructures are from 10 to 30 feet high and provide space beneath the derrick floor for the blowout preventer valves.

SUCKER RODS
Steel rods that are screwed together to form a "string" that connects the pump inside a well's tubing down hole to the pumping jack on the surface; pumping rods.

SUCKER RODS, HOLLOW
In certain applications, slim-hole pumping, hollow sucker rods are used, serving the dual purpose of rod and production tubing in the same string. Traveling-barrel pumps (q.v.) are most often used with hollow-rod pumping. The rods are attached to the cage or pull-tube (traveling barrel); the pump is installed in the seating nipple, or a packer-type pump anchor is used.

SUITCASE ROCK
Any formation that indicates further drilling is impractical. Upon hitting such a formation, drilling crews traditionally pack their suitcases and move on to another site.

SUPERCHARGER
A mechanism such as a blower or compressor for increasing the volume of air-charge to an engine over that which can normally be drawn into the cylinders through the action of the pistons on the suction strokes. Superchargers are operated or powered by an exhaust-gas turbine in the engine's exhaust stream.

SUPERCHARGE, TO
To supply air to an engine's intake or suction valves at a pressure higher than the surrounding atmosphere. *See* Supercharger.

SUPERPORT
A terminal or oil-handling facility located offshore in water deep enough to accommodate the largest, deep-draft oil tankers.

SURFACE PIPE
See Casing

SURGE TANK
A vessel on a flow line whose function is to receive and neutralize sudden, transient rises or surges in the stream of liquid. Surge tanks often are used on systems where fluids flow by heads (q.v.) owing to entrained gas.

SURVEY STAKES
Wooden markers driven into the earth by a survey crew identifying the boundaries of a right-of-way, the route of a pipeline, or a well location. Survey stakes may bear notations indicating elevation or location.

SUSPENDED DISCOVERY
An oil or gas field that has been identified by a discovery well but is yet to be developed.

SWAB, TO
(1) To clean out the bore hole of a well with a special tool attached to a wire line. Swabbing a well is often done to start it flowing. By evacuating the fluid contents of the hole the hydrostatic head is reduced sufficiently to permit the oil in the formation to flow into the bore hole, and if there is enough gas in solution the well may flow for a time. (2) To glean as much information as possible from a person; to pump someone for facts he would not normally volunteer.

SWAG
A downward bend put in a pipeline to conform to a dip in the surface of the right-of-way, or to the contours of a ravine or creek; a sag.

SWAGE
A heavy, steel tool, tapered at one end, used to force open casing that has collapsed down hole in a well.

SWAGE NIPPLE
An adapter; a short pipe fitting, a nipple, that is a different size on each end, e.g. 2-inch to 3-inch; 2-inch to 4-inch.

SWAMPER
A helper; the person who assists a truck driver load and unload and helps take care of the vehicle.

SWAY BRACES
The diagonal support-braces on a rig structure. Along with the horizontal girts, sway braces hold the legs (the corner members) of the rig in place.

SWEET
Having a good odor; a product testing negative to the "doctor test"—free of sulfur compounds.

SWEET CRUDE
Crude oil containing very little sulfur and having a good odor.

SWEET GAS
Natural gas free of significant amounts of hydrogen sulfide (H_2S) when produced.

SWING CHECK
A check valve (q.v.).

SWING JOINT
A combination of pipe fittings that permits a limited amount of movement in the connection without straining the lines, flanges, and valves.

SWING LINE
A suction line inside a storage or a working tank that can be raised or lowered by a wire line attached to a hand winch mounted on the outside of the tank. By raising the swing line above the level of water and sediment in the tank, only the clean oil is pumped out.

SWITCHER
A person who works on an oil lease overseeing the filling of lease stock tanks. When a tank is full he switches valves, turning the production into other tanks. A switcher works on a lease with flowing production; if the lease had only pumping wells, he would be called a pumper.

SWIVEL
A part of the well-drilling system; a heavy, steel casting equipped with a bail—held by the hook of the traveling block—containing the wash pipe, goose neck, and bearings on which the kelly joint hangs and rotates; the heavy link between the hook and the drill string onto which the mudhose is attached; an item of "traveling" equipment.

SYNCLINE
A bowl-shaped geological structure usually not favorable to the accumulation of oil and gas. Stratigraphic traps (q.v.) are sometimes encountered in synclines. See Anticline.

SYNFUEL
Short for synthetic gas or oil (q.v.).

SYNTHANE PLANT
A coal-to-gas pilot plant operated by the Energy Research and Development Administration in Pennsylvania. Designed to produce 1.2 MMCFD of pipeline gas, designated as synthane, synthetic methane.

SYNTHETIC GAS
Commercial gas made by the reduction or gasification of solid hydrocarbons: coal, oil shale and tar sand. See Gasification.

SYNTHETIC OIL
A term applied to oil recovered from coal, oil shales, and tar sands (q.v.).

T

TACK WELD
Spot welds temporarily joining two joints of pipe to hold them in position for complete welding.

TAIL, TO
To carry the light end of a load; to extricate a vehicle from a ditch or mud.

TAIL CHAIN
The short length of chain, with a hook attached, on the end of a winch line.

TAIL GAS
Residue gas from a sulfur recovery unit; any gas from a processing unit treated as residue.

TAILING — OUT RODS
Unscrewing and stacking rods horizontally outside the derrick. As a rod is unscrewed, a worker takes the free end and, as the elevators holding the other end is slacked off, he "walks" the rod to a rack where it is laid down.

TAILINGS
Leftovers from a refining process; refuse material separated as residue.

TANDEM, A
A heavy-duty, flat-bed truck with two closely coupled pairs of axles in the rear; a ten-wheeler.

TANK
(1) Cylindrical vessel for holding, measuring, or transporting liquids. (2) Colloquial for small pond; stock tank.

TANK BATTERY
See Battery

TANK BOTTOMS
Oil-water emulsion mixed with free water and other foreign matter that collect in the bottoms of stock tanks and large crude storage tanks. Periodically, tank bottoms are cleaned out by physically removing the material or by the use of chemicals which separate oil from water permitting both to be pumped out.

TANK DIKE
A mound of earth surrounding an oil tank to contain the oil in the event of a rupture in the tank, a fire, or the tank running over.

TANKER TERMINAL
A jetty or pier equipped to load and unload oil tankers. *See* Sea terminal.

TANK FARM
A group of large riveted or welded tanks for storage of crude oil or product. Large tank farms cover several square miles.

TANK MIXER
Motor-driven propeller installed on the shell of a storage tank to stir up and mix tank sediments with the crude. The propeller-shaft protrudes through the shell, with the motor mounted on the outside. Turbulence created by the prop thrust causes the BS&W to remain suspended in the oil as it is pumped out.

TANK TABLES
A printed table showing the capacity in barrels for each one-eighth inch or one-quarter inch of tank height, from bottom to the top gauge point of the tank. Tank tables are made from dimensions furnished by tank strappings (q.v.). *See* Strapping.

TANK TRAIN
A new concept in the rail shipment of crude oil, products, and other liquids developed by General American Transportation (GATX). "Tank Train" tank cars are interconnected which permits loading and unloading of the entire train of cars from one trackside connection. This arrangement does away with the need for the conventional loading rack (q.v.), and vapors from the filling operation can be more easily contained. *See* Densmore, Amos.

TAP SADDLE
A type of pipeline clamp with a threaded hole in one of the two halves of the bolt-on clamp for use when a pipeline is to be tapped, to have a hole made in it for drawing off gas or liquid. Tap saddles are used on field lines, 2″ to 10″; for tapping larger lines, nipples are welded to the pipe and a tapping machine (q.v.) is used.

TAPPED OR FLANGED CONNECTIONS
Indicates the two types of pump or process unit connections available from suppliers. Tapped is an internally threaded (female) connection into which an externally threaded piece may be screwed; a flanged connection is one furnished with a screw or weld flange.

TAPPING AND PLUGGING MACHINE
A device used for cutting a hole in a pipeline under pressure. A nipple, with a full-opening valve attached, is welded to the line. The tapping machine is screwed onto the valve and, working through the open valve, bores a hole in the line. The tapping drill is withdrawn, the valve is closed, and the tapping machine is unscrewed from the valve. A connection can then be made to the pipeline at the valve.

TAPS
Trans-Alaska Pipeline System; a large-diameter pipeline built from the oil-rich North Slope of Alaska to the warm-water port of Valdez on the state's south shore.

TARIFF
A schedule of rates or charges permitted a common carrier or utility; pipeline tariffs are the charges made by common carrier pipelines for moving crude oil or products.

TAR SANDS
See Petroleum tar sands

TATTLE TALE
A geolograph; a device to record the drilling rate or rate of penetration during an eight-hour tour (q.v.).

TBA
Among marketing department people, TBA stands for tires, batteries, and accessories.

T.D.
Total depth. Said of a well drilled to the depth intended.

TEAMING CONTRACTOR
A person who furnished teams of horses and mules and oil field wagons for construction and earth-work in the oil fields. Some large teaming contractors in the early days kept stables with 600 teams (1,200 horses and mules). In the days of dirt roads in the booming oil fields, the horse and wagon was the most dependable mode of transportation.

TEAMSTER
See Mule skinner

TEAPOT DOME
Part of the Naval Petroleum Reserves set aside by Congress in 1923. Teapot Dome in Wyoming was the center of controversy and scandal in the 1920s during the presidency of Warren G. Harding.

TECTONIC MAP
A geological map; a structural map showing the folding and faulting of subsurface formations.

TELEGRAPH
A device for the "remote" control of a steam drilling engine on a cabletool rig. The "telegraph" consisted of a wire or a small cable running between the pulleys, one at the driller's stand, the other mounted on the steam valve of the engine. By turning his wheel, the driller regulated the speed of the engine by opening or closing the steam valve.

TEMPERATURE BOMB
A device used down hole to measure bottom-hole and circulating temperatures on a drilling well. One technique involves attaching a temperature sensitive probe in a protective sleeve attached to a carrier mounted on the drill pipe.

TEMPERATURE CONVERSION
(F° to C°) C° = 5/9 (F° − 32°); C° to F°) F° = 9/5 (C° + 32°)

TEMPERATURE LOG
Recording temperature variations down hole by the use of an electrode containing a length of platinum wire that readily assumes the temperature of the drilling mud, gas, or water leaking into the hole. One important use of the logging device is to determine the location of cement in the annular space between casing and well bore after a cement job. The curing or hardening cement gives off heat which alters the flow of electric current observable at the surface.

TEMPER SCREW
A device on the cable of a string of cable tools that permits the driller to adjust the tension on the drilling line. A temper screw is made in the general form of a turnbuckle (q.v.).

TEMPLATE PLATFORM
An offshore platform whose supporting legs fit into a frame previously constructed and anchored to the sea floor. The platform, constructed on shore, is taken out to location by a crane-barge where it is set into the frame.

TENDER
(1) A permit issued by a regulatory body or agency for the transportation of oil or gas. (2) A barge or small ship serving as a supply and storage facility for an offshore drilling rig; a supply ship.

TENDERS
A quantity of crude oil or refined product delivered to a pipeline for transportation. Regulations set the minimum amount of oil that will be accepted for transportation.

TENSIOMETER
A gauge attached to a cable or wire rope to detect the tension being applied. From two positions on a section of the cable a sensitive gauge measures the stretch and twist of the cable under load, indicating the tension on a scale; a strain gauge.

TENSIONER SYSTEMS
Tensioner systems are installed on deep-water floating drilling platforms to maintain a constant tension on the marine-riser (q.v.). Two types of

systems are used; the deadweight system and the pneumatic system. Tensioning systems serve the dual purpose of compensating for the vertical motion of the drilling vessel or platform and maintaining a constant tension or lifting force on the riser.

TENSION-LEG PLATFORM
A semisubmersible drilling platform held in position by multiple cables anchored to the ocean floor. The constant tension of the cables makes the platform immune to heave, pitch, and roll caused by wave action, conditions that affect conventional semisubmersibles.

TERTIARY RECOVERY
The third major phase of crude oil recovery. The primary phase is flowing and finally pumping down the reservoir until it is "depleted" or no longer economical to operate. Secondary recovery usually involves repressuring or simple water flooding. The third or tertiary phase employs more sophisticated techniques of altering one or more of the properties of crude oil, e.g. reducing surface tension. This is accomplished by flooding the formations with water mixed with certain chemicals that "free" the oil adhering to the porous rock so it may be taken into solution and pumped to the surface. *See* Micellar/surfactant flooding.

TEST COUPONS
Small samples of materials—metals, alloys, coatings, plastics and ceramics—which are subjected to heat, cold, pressure, humidity and other conditions of stress to test durability and performance under simulated operating conditions.

TEST SET (TELEPHONE)
A lineman's portable equipment for testing the circuit on a telephone line. The test set includes a hand-cranked telephone instrument whose lead wires are clipped to the phone lines when the lineman wants to call in to the switchboard.

TETRAETHYL LEAD
A lead compound added, in small amounts, to gasoline to improve its anti-knock quality. Tetraethyl lead (TEL) is manufactured from ethyl chloride which is derived from ethylene, a petrochemical gas.

TEXAS DECK
The top deck of a large semi-submersible drilling platform. The upper deck of any offshore drilling rig that has two or more platform levels.

TEXAS TOWER
A radar or microwave platform supported on caissons anchored to the ocean floor. The tower resembles an offshore drilling platform in the Texas Gulf, hence the name.

THERMAL CRACKING
A refining process in which heat and pressure are used to break down, rearrange, or combine hydrocarbon molecules. Thermal cracking is used to increase the yield of gasoline obtainable from crude oil.

THERMAL OXIDIZERS
A large, cylindrical furnace, with refractory lining and banks of burners at various levels, for burning refinery gases before they are vented to the flare tower (q.v.).

THERMOCOUPLE
A pyrometer; a temperature-measuring device used extensively in refining. The thermocouple is based upon the principle that a small electric current will flow through two dissimilar wires properly welded together at the ends, when one junction is at a higher temperature than the other. The welded ends are known as the "hot junction" which is placed where the temperature is to be measured. The two free ends are carried through leads to the electromotive force detector, known as the "cold junction." When the hot junction is heated, the millivolts can be measured on a temperature scale.

THERMOMETRIC HYDROMETER
A hydrometer (q.v.) which has a thermometer as an integral part of the instrument to show the temperature of the liquid. This is of first importance as the density or API gravity varies with the temperature. Hydrometers used by pipeline gaugers are thermometric hydrometers.

THIEF
A metal or glass cylinder with a spring actuated closing device that is lowered into a tank to obtain a sample of oil, or to the bottom of the tank to take a column of heavy sediment. The thief is lowered into the tank on a line that when jerked will trip the spring-valve enabling the operator to obtain a sample at any desired level.

THIEF HATCH
An opening in the top of a tank large enough to admit a thief and other oil-sampling equipment.

THIEFING A TANK
Taking samples of oil from different levels in a tank of crude oil, and from the bottom to determine the presence of sediment and water with the use of a thief (q.v.).

THIEF ZONE
A very porous formation down hole into which drilling mud is lost. Thief zones, which also include crevices and caverns, must be sealed off with a liner or plugged with special cements or fibrous clogging agents before drilling can resume.

THIRD-GENERATION HARDWARE
Equipment developed from earlier, less sophisticated models or prototypes; the latest in the evolution of specialized equipment.

THIXOTROPIC
The property of certain specially formulated cement slurries—used in cementing jobs downhole—that causes them to "set," become rigid when pumping ceases. But when force is again applied (pumping is resumed) the cement again becomes a pumpable slurry. This procedure may be repeated until the predetermined setting-time of the cement is reached.

THREAD PROTECTOR
A threaded cap or light-weight collar screwed onto the ends of tubular goods (pipe, casing, and tubing) to protect the threads from damage as the pipe is being handled.

THRIBBLES
Drill pipe and tubing pulled from the well three joints at a time. The three joints make a stand (q.v.) of pipe that is racked in the derrick. Two-joint stands are doubles; four-joint stands are fourbles.

THROWING THE CHAIN
Wrapping the spinning chain (q.v.) around the drill pipe in preparation for running the pipe up or backing it out. Crew members become proficient at throwing the chain in such a way as to put several wraps on the pipe with one deft motion.

THRUSTERS
Jets or propellers on large tankers, drill ships, and deepwater drilling platforms that provide a means to move the vessel sideways—at right angles to the ship's normal line of travel—when docking or in maintaining position in water too deep for conventional anchors. See Dynamic station keeping.

THUMPER
See Vibrator vehicle

TIDELANDS
Land submerged during high tide. The term also refers to that portion of the continental shelf between the shore and the boundaries claimed by the states. The Federal government now has the right to produce oil and gas from this area of the Continental Shelf.

TIE-IN
An operation in pipeline construction in which two sections of line are connected: a loop tied into the main line; a lateral line to a trunk line.

TIGHT HOLE
A drilling well about which all information—depth, formations encountered, drilling rate, logs—is kept secret by the operator.

TIN HAT
The metal, derby-like safety hat worn by all workers in the oil fields, refineries, and plants to protect their heads.

TLP
Term-limit pricing; an agreement on price between a supplier and a wholesaler or jobber that runs for a specified length of time.

TOE BOARD
The enclosure at toe height around a platform or on a catwalk ro prevent tools or other objects on the platforms from being kicked off accidentally.

TOOL DRESSER
In cable tool drilling, a worker who puts a new cutting edge on a drill bit that is worn or blunted. Like a blacksmith, the tool dresser heats the bit in a charcoal fire and using a hammer draws out the metal into a sharp, chisel-like cutting edge.

TOOLIE
A tool dresser (q.v.) on a cable-tool rig.

TOOL JOINTS
Heavy-duty, threaded joints specially designed to couple and uncouple drill pipe into "stands" (q.v.) of such length that they can be racked in the derrick. Intermediate couplings between the tool joints are made with regular drill pipe collars.

TOOL PUSHER
A supervisor of drilling operations in the field. A tool pusher may have one drilling well or several under his direct supervision. Drillers are directed in their work by the tool pusher.

TOP OUT
To finish filling a tank; to put in an additional amount that will fill the tank to the top.

TOPPED CRUDE OIL
Oil from which the light ends (q.v.) have been removed by a simple refining process.

TOPPING PLANT
An oil refinery designed to remove and finish only the lighter constituents of crude oil, such as gasoline and kerosine. In such a plant, the oil remaining after these products are taken off is usually sold as fuel oil.

TOPS
The "tops" in a refinery operation are the fractions or products distilled or flashed off at the top of a tower or distillation unit.

TORPEDO
An explosive device used in shooting (q.v.) a well. The well-shooting torpedo was invented and used by Col. E. A. L. Roberts, a Civil War veteran, in 1865. The first torpedoes used black powder as an explosive; later, nitroglycerin was substituted for the powder.

TORQUE
A turning or twisting force; a force that produces a rotation or torsion, or tends to.

TORQUE WRENCH
A tool for applying a turning or twisting motion to nuts, bolts, pipe or anything to be turned and which is equipped with a gauge to indicate the force or torque being applied. Torque wrenches are useful in tightening a series of bolts or nuts with equal tension, as on a flange or engine head.

TORREY CANYON
An oil tanker that ran aground off the coast of England causing the largest, most costly, and most publicized oil spill up to that time. The mishap touched off reactions that put oil-spill pollution in the international spotlight. The Torrey Canyon ran aground March 18, 1967.

TORSION BALANCE
A delicate instrument used by early-day geophysical crews to measure the minute variations in magnetic attraction of subsurface rock formations. As the differences in attraction of the subsurface features were plotted over a wide area, the geophysicist had some idea as to where sedimentary formations that might contain oil were located in relation to non-sedimentary rocks. The torsion balance has been superceded by the less complicated (to use) gravity meter or gravimeter. *See* Gravimeter and Magnetometer.

TOUR
A work period; a shift of work, usually eight hours, performed by drilling crews, pump station operators, and other oil field personnel. In the field, "tour" is pronounced tower, to rhyme with sour; a trick.

TOWER HAND
A member of the drilling crew who works up in the derrick; derrick man.

TRACER LINES
Small-diameter tubing paralleling and in contact with process or instrumentation piping in a refinery or other plant to provide heat or cooling for

the fluid or gases in transit. More often tracer lines carry steam. In the field, larger diameter tracer lines are used to heat low-gravity, viscous crude oils so they may be pumped. *See* also Heat tape.

TRACTOR FUEL
A low-octane fuel, less volatile than motor gasoline, used in low-compression farm tractors.

TRADER
One who deals in bulk petroleum or products both domestic and foreign; one who operates in the international oil market, arranging for supplies and trading surpluses of one product for others; an oil broker.

TRANSITION FITTINGS
When using plastic pipe in the field or at a plant, it is usually necessary to make connection with steel tank fittings or a pipeline. If so, special transition fittings, made with one end acceptable to the plastic pipe, the other end a standard thread-end or weld-end are installed.

TRANSSHIPMENT TERMINAL
A large, deep-water terminal where crude oil and products are delivered by "supertankers" (LCCV) (q.v.) and transshipment of product is by smaller tankers. Such terminals have large storage capacities and high-volume unloading facilities to accommodate the mammoth vessels that carry more than two million barrels of oil each trip.

TRAVELING BLOCK
The large, heavy-duty block hanging in the derrick and to which the hook is attached. The traveling block supports the drill column and "travels" up and down as it hoists the pipe out of the hole and lowers it in. The traveling block may contain from three to six sheaves depending upon the loads to be handled and the mechanical advantage necessary. The wire line from the hoisting drum on the draw works runs to the derrick's crown block and down to the traveling block's sheaves.

TREATING PLANT
A facility for heating oil containing water, emulsions, and other impurities and with the addition of chemicals causing the water and oil to separate. The water and other foreign matter settle to the bottom of the tank and is then drawn off.

TRICK
See Tour

TRICKLE-CHARGED BATTERY
A storage battery, usually for standby, emergency service, kept charged by a small amount of current from a primary electrical source. Should the main source of power fail, the battery, fully charged, is ready for use.

TRIP
See Round trip

TRIPLEX PUMP
A reciprocating pump with three plungers or pistons working in three cylinders. The triplex pump discharges fluid more evenly than a duplex or two-plunger pump, as it has a power stroke every one-third of a revolution of the crankshaft compared to every half revolution for the duplex pump.

TRIPPING THE BIT
Removing the bit from the hole and running it in again. (In removing the bit, the drill pipe must be pulled a stand at a time in order to reach the bit.) *See* Round trip.

TRUNK LINE
Main line (q.v.).

TUBE BUNDLE
The name given to the tubes in the core of a heat exchanger (q.v.). The tubes or pipes, all the same length, are spaced equidistance apart in parallel rows and are supported by perforated endplates thus forming a "bundle."

TUBE STILL
A pipe still (q.v.).

TUBE TURN
A weld or flanged fitting in the shape of a U used in construction of manifolds, exchanger bundles, and other close pipe work.

TUBING ANCHOR
A downhole, packer-like device run in a string of tubing that clamps against the wall of the casing. The tubing anchor prevents the "breathing" of the tubing, the cyclic up and down movement of the lower section of tubing as the well is pumped by a rod pump.

TUBING AND CASING ROLLERS
A downhole tool for reconditioning buckled, dented or collapsed well tubing or casing. The tool is lowered into the hole, entering the small, deformed diameter of the damaged pipe. As the cylindrical tool is forced lower and rotated it pushes out dents and restores the pipe to its original diameter.

TUBING BOARD
A small platform high in the derrick where a "derrick man" (a member of the drilling crews who is not affected with acrophobia) stands to rack drill pipe or tubing as it is being pulled and set back (q.v.).

TUBING HEAD
The top of the string of tubing with control and flow valves attached. Similar in design and function to the casinghead, the tubing head supports the string of tubing in the well, seals off pressure between casing and the inside of the tubing, and provides connections at the surface to control the production of gas or oil.

TUBING PUMPS
See Pumps, Tubing

TUBULAR GOODS
Refers to drill pipe, casing, well tubing, and line pipe; a generic term for any steel pipe used in the oil fields.

TURBOCHARGER
A centrifugal blower driven by an engine's exhaust-gas turbine to supercharge the engine. To supercharge (q.v.) is to supply air to the intake of an engine at a pressure higher than the surrounding atmosphere.

TURBULENT FLOW
The movement of liquid through a pipeline in eddies and swirls which tends to keep the column of liquid "together" rather than running like a river with the center of the stream moving faster than the edges. *See* Plastic flow.

TURNAROUND
The planned, periodic inspection and overhaul of the units of a refinery or processing plant; the preventive maintenance and safety check requiring the shutting down of a refinery and the cleaning, inspection, and repair of piping and process vessels.

TURNBUCKLE
A link with a screw thread at one end and a swivel at the other; a right-and-left screw link used for tightening a rod, a guy wire, or stay.

TURNKEY CONTRACT
A contract in which a drilling contractor agrees to furnish all materials and labor and do all that is required to drill and complete a well in a workmanlike manner. When on production, he "delivers" it to the owner ready to "turn the key" and start the oil running into the lease tank, all for an amount specified in the contract.

TURNKEY, TO
A verb made from the adjective "turnkey;" to perform a complete job as under a turnkey contract (q.v.); to take over and perform all necessary work of planning, procurement, construction, completion, and testing of a project before turning it over to the owner for operation.

TURNKEY WELL
A well drilled under a turnkey contract (q.v.).

TURNTABLE, ROTARY DRILLING
See Rotary table

TURTLEBACK
A two-part clamp for joining lengths of shackle rod (q.v.). The connector is in the general configuration of an English walnut; the two halves are held together by a bolt and nut.

TVD
Total vertical depth. TVD is always less than a well's total depth (TD) because of the inevitable deviation from the vertical of the well bore.

TWIST A TAIL
To bring pressure to bear in order to speed up a job or to get action from someone who is suspected of dragging his feet.

TWO-CYCLE ENGINE
An internal combustion engine that produces one power stroke for each revolution of the crankshaft. Intake, compression, ignition, and power stroke are accomplished in one revolution.

TWO-STAGE COMPRESSOR
Two-stage identifies a type of compressor that intakes gas and compresses or raises the pressure in the first chamber of the compressor and passes the gas into the second-stage chamber where it is further compressed, raising the pressure to the required level.

U

U-BOLT
A bolt in the shape of a U, both ends of which are threaded. A follower or saddle-piece fits over the threaded ends and held in place by nuts. U-bolts or U-clamps are used to hold two ends of wire lines together or to make a loop in a length of wire cable by turning back the running part (the loose end) on the standing part of the cable and clamping them together.

ULTRASONIC ATOMIZER
A development in burners for heating oils in which high frequency sound waves are focused on the stream of fuel, forming a spray of microscopic

fuel droplets. The resulting intimate mixture of fuel and air makes for greater combustion efficiency.

UNASSOCIATED GAS
Natural gas occurring alone, not in solution or as free gas with oil or condensate. *See* Associated gas.

UNBRANDED GASOLINE
Gasoline sold by major refiners to jobbers and other large distributors without bearing the name of the refiner.

UNDERGROUND STORAGE
In certain areas of the country where there are underground caverns petroleum and products are stored for future use. All caverns are not suitable; some are not naturally sealed and would permit the stored oil to leak into subsurface water sources. *See* Salt-dome storage.

UNDERREAM, TO
To enlarge the size of the bore hole of the well by the use of an under-reamer (q.v.), a tool with expanding arms or lugs that, when lowered into the hole, can be released at any depth to ream the hole with steel or insert cutters.

UNDERREAMER
A type of drilling tool used to enlarge the diameter of the bore hole in certain downhole intervals. The underreamer is made with expandable arms fitted with cutters. When in position the expandable arms are released and the cutters chew away the rock to enlarge the hole. When the reamer is pulled from the hole the arms fold in toward the body of the tool.

UNITIZATION
A term denoting the joint operation of separately owned producing leases in a pool or reservoir. Unitization makes it economically feasible to undertake cycling, pressure maintenance, or secondary recovery programs.

UNIT OPERATOR
Head well-puller; the man in charge of the pulling unit crew that does routine subsurface work on producing wells, e.g. cleaning-out, changing pumps, pulling rods and tubing.

UNMANNED STATION
A pipeline pumping station that is started, stopped, and monitored by remote control. Through telecommunication systems, most intermediate booster stations on large trunk lines are unmanned and remotely controlled from the dispatcher's office.

UNIVERSAL JOINT
A shaft coupling able to transmit rotation to another shaft not directly in line

with the first shaft; a moveable coupling for transmitting power from one shaft to another when one shaft is at an angle to the other's long axis.

UP-DIP WELL
A well located high on a structure where the oil-bearing formation is found at a shallower depth.

UPSET TUBING
Tubular goods that are "upset" are made thicker in the area of the threads in order to compensate for the metal cut away in making threads. In the manufacture of casing and drill pipe, the additional metal is usually put on the inside, but in well tubing, especially the smaller sizes, the thickening is on the outside. This is known as exterior-upset tubing.

U.S.G.S.
U.S. Geological Survey; an agency of the Federal government that, among its many services and duties, regulates the placement of wells in Federal offshore lands.

V

VACUUM DISTILLATION
Distillation under reduced pressure (less than atmospheric) which lowers the boiling temperature of the liquid being distilled. This technique with its relatively low temperatures prevents cracking or decomposition of the charge stock (q.v.).

VACUUM STILL
A refining vessel in which crude oil or other feedstock is distilled at less than atmospheric pressure.

VACUUM TAR
See Asphalt

VALVE POTS
The wells in the body of a reciprocating (plunger) pump where the suction and discharge valves are located. Valve pots are on the fluid end of the pump, and are covered and sealed by heavy, threaded plugs or metal caps bolted over the top of the pots.

VANE PUMP
A type of rotary pump designed to handle relatively small volumes of liquid products: gasoline and light oils as well as highly viscous fluids.

VAN SYCKEL, SAMUEL

The man who invented and successfully operated the first crude oil pipeline. The line was two-inch and ran from Pithole City, Pa., to a railhead five miles away. It pumped 81 barrels the first day, thus sounding the knell for the teamster and his wagon load of oil barrels.

VAPOR LOCK

A condition that exists when a volatile fuel vaporizes in an engine's fuel line or carburetor preventing the normal flow of liquid fuel to the engine. To handle gas lock or vapor lock the gas must be bled off the system by removing a line or loosening a connection, or the lines and carburetor cooled sufficiently to condense the gas back to a liquid.

VAPOR PRESSURE

The pressure exerted by a vapor held in equilibrium with its liquid state. Stated inversely, it is the pressure required to prevent a liquid from changing to a vapor. The vapor pressure of volatile liquids is commonly expressed in pounds per square inch absolute (q.v.) at a temperature of 100° F. For example, butane has a vapor pressure of 52 p.s.i.a. at 100° F.

VAPOR RECOVERY UNIT

A facility for collecting and condensing vapors of volatile products being loaded into open tanks at refineries, terminals, and service stations. The vapors are drawn into a collecting tank and by pressure and cooling are condensed to a liquid. VR units significantly reduce air pollution by petroleum vapors.

VAPOR TENSION

See Vapor pressure

V-BELT

A type of "endless" V-shaped belt used in transmitting power from an engine's grooved drive-pulley to the grooved sheave of a pump, compressor, or other equipment. The V-belt, a bigger and tougher version of the automobile fan belt, is used in sets of from two to twenty belts depending upon the size of the drive-pulley.

V-DOOR

The opening in the derrick opposite the draw works used for bringing in drill pipe and casing from the nearby pipe racks.

VENTURI METER

An instrument for measuring the volume of flowing gases and liquids. It consists of two parts—the tube through which the fluid flows and a set of indicators which show the pressures, rate of flow, or quantity discharged. The tube, in the shape of an elongated hour glass, is flanged into a pipeline carrying the fluid. The effect of the tube is to increase the velocity and decrease the pressure at the point where the tube's diameter is reduced. The relationship between the line pressure and the pressure at the narrow "waist" of the tube is used in computing the rate of flow.

VERTICAL INTEGRATION
Refers to the condition in which a company produces raw material, transports it, refines or processes it, and markets the product, all as one integrated operation. Specifically, an oil company is said to be vertically integrated when it finds and produces oil and gas; transports it in its own pipelines; refines it; and markets its products under its brand name. According to the critics of the industry, this is not in the country's best interest. See also Horizontal integration.

VIBRATOR VEHICLE
A specially designed tractor-like vehicle used to produce shock waves for geophysical and seismic surveys. The vehicle incorporates a hydraulically operated hammer or "thumper" that strikes the ground setting off shock waves which are reflected from subsurface rock formations and recorded by seismic instruments at the surface.

VISCOSITY
One of the physical properties of a liquid, i.e., its ability to flow. It happens that the more viscous an oil, for example, the less readily it will flow, so the term has an inverse meaning—the lower the viscosity, the faster the oil will flow. Motor oil with a viscosity of SAE 10 flows more readily than oil with a viscosity of SAE 20. See Seconds Saybolt.

VISCOSITY INDEX
An arbitrary scale used to show the changes in the viscosity of lubricating oils with changes in temperatures.

VLCC
Very large crude carrier; a crude oil tanker of 160,000 deadweight tons or larger, capable of transporting one million barrels or more.

VLPC
Very large product carriers (ocean-going tankers).

V.O.I.C.E.
Voluntary Oil Industry Communications Effort; part of a full-scale advertising/information program conducted by the American Petroleum Institute to tell the oil industry's story to the public. V.O.I.C.E. or VOICE is part of the program in which speakers from the industry appear before interested groups to tell oil's story.

VOLATILITY
The extent to which gasoline or oil vaporizes; the ease with which a liquid is converted into a gaseous state.

VOLUME TANK
A small cylindrical vessel connected to a gas line in the oil field to provide an even flow of gas to an engine, and to trap liquids that may have condensed in the gas line.

W

WALKING BEAM
A heavy timber or steel beam supported by the Samson post, that transmits power from the drilling engine via the bandwheel, crank, and pitman (q.v.) to the drilling tools. The walking beam rocks or oscillates on the Samson post imparting an up-and-down motion to the drilling line or to the pump rods of a well.

WALL CAKE
See Filter cake

WALL CLEANER
A scraping or cutting device attached to the lower joints of casing in the string for the purpose of cleaning the wall of the bore hole in preparation for cementing. There are numerous types of scratching, raking, and cutting devices designed to remove the clay sheet or "filter cake" deposited by the circulating drilling mud. Mechanically cleaning the walls frees the production formation from the caked mud and also enlarges the hole diameter through the production zone making for more efficient oil flow into the well bore.

WALL SCRAPER
See Wall cleaner

WALL STICKING (OF DRILL PIPE)
A condition down hole when a section of the drill string becomes stuck or hungup in the deposit of filter cake on the wall of the bore hole. Also referred to as "differential sticking," an engineer's term.

WARM UP
To hammer a pipe coupling so as to loosen the threaded connection. Repeated pounding with a hammer literally warms the connection as well as "shocking" corroded threads so that they can be unscrewed.

WASTING ASSET
A material (usually mineral property) whose use results in depletion; a non-replaceable mineral asset; oil, gas, coal, uranium, and sand.

WATER-CONING
The encroachment of water in a well bore in a water-drive reservoir owing to an excessive rate of production. The water below the oil moves upward to the well bore through channels, fissures, and permeable streaks leaving the oil sidetracked and by-passed.

WATER DRIVE
The force of water under immense pressure below the oil formation that, when the pressure is released by drilling, drives the oil to surface through the well bore.

WATER-DRIVE RESERVOIR
An oil reservoir or field in which the primary natural energy for the production of oil is from edge- or bottom-water in the reservoir. Although water is only slightly compressible, the expansion of vast volumes of it beneath the oil in the reservoir will force the oil to the well bore.

WATER FLOODING
One method of secondary recovery (q.v.) in which water is injected into an oil reservoir to force additional oil out of the reservoir rock and into the well bores of producing wells.

WATER-LOSS IN DRILLING MUD
See Filtration-loss quality of mud

WATER PRESSURE
Pounds per square inch = height of the water column in feet X 0.434; e.g. 10-foot column of water X 0.434 = 4.34 pounds per square inch of pressure.

WATER TABLE
The level of ground water in the earth; the surface below which all pores of the ground are filled with water.

WEATHERED CRUDE
Crude oil which has lost an appreciable quantity of its volatile components owing to natural causes (evaporation) during storage and handling.

WEATHERING
The old practice of allowing highly volatile products such as natural gasoline (q.v.) to stand in tanks vented to the atmosphere to "weather off," to lose some of the more volatile fractions before being pumped into a pipeline. This wasteful procedure is no longer permitted.

WEEVIL-PROOF
Refers to tools and equipment simple to operate or to assemble; fittings and equipment parts impossible for a boll weevil, a green hand, to put together improperly.

WEIGHBRIDGE

A facility to measure the contents of rail tank cars loading LP-gas at a refinery or terminal. The rail cars are moved onto the scales and loading is done while the cars are on the weighbridge (scales). When the tank car is filled, the flow is automatically shut off and a ticket for the net weight is printed simultaneously.

WELD-END FITTINGS

Nipples, flanges, valves, and plugs without threads, but with plain, beveled ends that can be welded to non-threaded, plain-end pipe. For proper welding, the ends of both fittings and pipe are beveled to provide a V-shaped groove for the courses of welding metal.

WELDER'S HELPER

A person who assists the welder. His most important job is to brush the weld with a wire brush to dislodge rust and scale. He also keeps the welder supplied with rods and holds or turns the piece being welded.

WELDING BOTTLES

Steel cylinders of oxygen and acetylene gas used in oxyacetylene or gas welding. The oxygen bottle is green and taller and smaller in diameter than the black acetylene bottles.

WELDING-BOTTLE GAUGES

A type of small, adjustable-flow regulators screwed onto oxygen and acetylene gas bottles to regulate the flow of gases to the welding torch (q.v.).

WELDING "BUG"

An automatic electric welding unit; specifically, the welding head that contains the welding-wire electrode and moves on the pipe's circumference on an aligned track-like guide. Used in welding large-diameter line pipe.

WELDING, CO_2 SHIELDED

A semi-automatic technique of field welding that has the advantage of making welding a hydrogen-free operation, thus eliminating hydrogen cracking of the weld metal; an inert gas-shielded welding process; electric welding in which the molten metal being laid down is blanketed by CO_2 to protect it from active gases making contact with the molten surface.

WELDING GOGGLES

Dark, safety glasses used by oxyacetylene welders and welders' helpers to protect their eyes from the intense light of the welding process and from the flying sparks. *See* Welding hood.

WELDING HOOD
A wrap-around face and head shield used by electric welders. The hood has a dark glass window in the face shield; the hood tilts up, pivoting on the headband.

WELDING MACHINE
A self-contained electric generating unit composed of a gasoline engine direct-connected to a D.C. generator which develops current for electric welding. Welding machines or units are skid-mounted and transportable by dragging or by truck.

WELDING TORCH
An instrument used to produce a hot flame for welding; a hand-held, tubular device connected by hoses to a supply of oxygen and acetylene and equipped with valves for regulating the flow of gases to the "tip" or welding nozzle. By opening the valves to permit a flow of the two gases from the tip, the torch is ignited and then with adjustments of the valves, a hot (3,500° F.) flame results.

WELD, WET
A weld made under water, "in the wet" without the use of a "dry-box" as in hyperbaric (extreme pressure) welding (q.v.).

WELL
A hole drilled or bored into the earth, usually cased with metal pipe, for the production of gas or oil. Also, a hole for the injection under pressure of water or gas into a subsurface rock formation. *See* Service well.

WELLHEAD CELLAR
An air-tight submarine chamber enclosing an underwater wellhead, large enough to permit work to be carried on in a dry and normal atmosphere. The wellhead cellar is a piece of equipment that has been developed for well completions and other work in a deep-water environment.

WELL NAMING
The naming of a well follows a long-standing, logical practice. First is the name of the operator or operators drilling the well; then the land owner from whom the lease was obtained; and last the number of the well on the lease or the block. For example Gulf Oil drills on a lease acquired from Dorothy Doe, and it is the first well on the lease. The name will appear in the trade journals as Gulf Doe No. 1, sometimes Gulf 1 Doe. A lease from the State of Oklahoma by Gulf and the name would appear on a sign at the well as: Gulf State No. 1 and perhaps followed by its location—SW NW 2 - 29s - 13w (SW quarter of NW quarter (40 acres) of Section 2, Township 29 south, Range 13 west.)

WELL PERMIT
The authorization to drill a well issued by a state regulatory agency.

WELL PLATFORM
An offshore structure with a platform above the surface of the water that supports the producing well's surface controls and flow piping. Well platforms often have primary separators and well-testing equipment. *See* Producing platform.

WELL PROGRAM
The step by step procedure for drilling, casing, and cementing a well. A well program includes all data necessary for the tool pusher (q.v.) and drilling crews to know: formations to be encountered, approximate depth to be drilled, hole sizes, bit types, sampling and coring instructions, casing sizes, and methods of completion—or abandonment if the well is dry.

WELL SHOOTER
A person who uses nitroglycerin and other explosives to shoot a well, to fracture a subsurface rock formation into which a well has been drilled. The shooter lowers the explosive into the well bore on a wire line. When the explosive charge has been landed at the proper depth, it is detonated electrically.

WELL STIMULATION
See Stimulation

WET GAS
Natural gas containing significant amounts of liquefiable hydrocarbons.

WET JOB
Pulling tubing full of oil or water. As each joint or stand (q.v.) is unscrewed, the contents of the pipe empties onto the derrick floor, drenching the roughnecks (q.v.). The tubing is standing full of fluid because the pump valve on the bottom of the tubing is holding and will not permit the fluid to drain out as it is being hoisted out of the hole.

WHIRLEY
The name applied to a full-revolving crane for offshore duty. Other barge-mounted cranes revolve 180°, over the stern and over both sides of the vessel.

WHIPSTOCK
A tool used at the bottom of the bore hole to change the direction of the drilling bit. The whipstock is, essentially, a wedge that crowds the bit to the side of the hole causing it to drill at an angle to the vertical.

WHITE, DR. ISRAEL CHARLES
The "father of petroleum geology." Dr. White brought about the transition from superstition and "creekology" to scientific geological methods. He was a poor West Virginia boy who grew up to become world famous as the discoverer of the anticlinal or structural theory of oil accumulation.

WHITE CARGO
Clean cargo; a term to describe distillate—gasoline, kerosine, heating oils—carried by tankers.

WHITTAKER SYSTEM
A patented system for protection of crews working on offshore drilling platforms, semi-submersibles, and other structures. The heart of the system is a survival capsule into which offshore crew members can retreat in the event of fire or explosion or other disaster and lower themselves to the water. The capsule is self-propelled and provides food, water, First Aid supplies, for 28 persons. Large offshore structures have several survival capsules that hang from davits at various locations on the platform. *See* Brucker survival capsule.

WICK OILER
A lubricator for large, slow-moving crank bearings. Oil is fed from a small canister, a drop at a time, onto a felt pad or "wick." As the crank turns beneath the wick, a scraper on the crank makes contact with the wick taking a small amount of oil.

WILDCATTER
A person or company that drills a wildcat well; a person held in high esteem by the industry, if he is otherwise worthy; an entrepreneur to whom taking financial risks to find oil is the name of the game.

WILDCAT WELL
A well drilled in an unproved area, far from a producing well; an exploratory well in the truest sense of the word; a well drilled out where the wildcats prowl and "the hoot owls mate with the chickens."

WINCH
A device used for pulling or hoisting by winding rope or cable around a power-driven drum or spool.

WINDLASS
A winch; a steam or electric-driven drum, with a vertical or horizontal shaft, for raising a ship's anchor.

WIRE LINE
A cable made of strands of steel wire; the "lines" used on a drilling rig.

WIRELINE TOOLS
Special tools or equipment made to be lowered into the well's bore hole on a wire line (small-diameter steel cable), e.g. logging tools, packers, swabs, measuring devices, etc.

WIRELINE TRUCK
A service vehicle on which the spool of wire line is mounted for use in down hole wireline work.

WITCHING
See Doodle bug

WOC TIME
Waiting-on-cement time; the period between the completion of the actual cementing operations and the drilling out of the hardened cement plug in the casing at the bottom of the well.

WOODCASE THERMOMETER
A thermometer used by gaugers in taking the "tank temperature," the temperature of the oil in the tank as contrasted to the temperature of the sample oil to be tested. The thermometer is encased in a wood frame to which a line may be attached for lowering the thermometer into the oil.

WORKING INTEREST
The operating interest under an oil and gas lease. The usual working interest consists of seven-eighths of the production subject to all the costs of drilling, completion, and operation of the lease. The other one-eighth of production is reserved for the lessor or landowner. *See* Landowner's royalty.

WORKING PRESSURE
The pressure at which a system or item of equipment is designed to operate. Normal pressure for a particular operation.

WORKING TANK
A terminal or main-line tank pumped into and out of regularly; a tank that is "worked" as contrasted to a storage tank not regularly filled and emptied.

WORKOVER
Operations on a producing well to restore or increase production. Tubing is pulled and the casing at the bottom of the well is pumped or washed free of sand that may have accumulated.

WORM GEAR
A type of pinion gear mounted on a shaft, the worm gear meshing with a ring gear; a gear in the shape of a continuous spiral or with the appearance of a pipe thread; often used to transmit power at right angles to the power shaft.

WRINKLE PIPE
To cut threads on a piece of pipe in order to make a connection.

WRIST PIN
The steel cylinder or pin connecting the rod to the engine's or pump's piston. The wrist pin is held in the apron or lower part of the piston by a friction fit and a circular, spring clip. The upper end of the connecting rod is fitted with a lubricated bushing which permits the rod to move on the pin. A piston pin.

X-Y-Z

XYLENE
An aromatic hydrocarbon; one of a group of organic hydrocarbons (benzene, toluene, and xylene) which form the basis for many synthetic organic chemicals. *See* BTX.

ZEOLITIC CATALYST
Catalyst formulations that contain zeolite (any of various hydrous silicates, a mineral) for use in catalytic cracking units.

ZONE
An interval of a subsurface formation containing one or more reservoirs; that portion of a formation of sufficient porosity and permeability to form an oil or gas reservoir.

ZONE ISOLATION
A method of sealing off, temporarily, a producing formation while the hole is being deepened. A special substance is forced into the formation where it hardens allowing time for the well bore to be taken on down. After a certain length of time, the substance again turns to a liquid unblocking the producing formation.

Bibliography

American Petroleum Institute, Division of Refining, GLOSSARY OF TERMS USED IN PETROLEUM REFINING, Washington, D.C., 1962.

——PETROLEUM—THE STORY OF AN AMERICAN INDUSTRY, Second Edition, 1935.

——GLOSSARY OF TERMS USED IN PETROLEUM REFINING, 1953

Ball, Max W., THIS FASCINATING OIL BUSINESS, New York, N.Y., Bobbs-Merrill, 1940.

Bland, W., and Davidson, PETROLEUM PROCESSING HANDBOOK, New York, N.Y., McGraw-Hill Publishing Co., 1967.

Bank of Scotland Information Service, OIL AND GAS INDUSTRY GLOSSARY OF TERMS, Edinburgh, Bank of Scotland Oil Div., 1974.

Brantly, J. E., ROTARY DRILLING HANDBOOK, New York, Palmer Publications, 1952.

Craft, B. C. and Hawkins, APPLIED PETROLEUM RESERVOIR ENGINEERING, Englewood Cliffs, N.J., Prentiss-Hall, 1959.

Desk and Derrick Clubs, D & D STANDARD ABBREVIATOR, Tulsa, Okla., Petroleum Publishing Company.

Fanning, Leonard M., MEN, MONEY & OIL: The Story of An American Industry, New York, N.Y., World Publishing Company, 1966.

Harris, L. M., DEEPWATER FLOATING DRILLING OPERATIONS, Tulsa, Oklahoma, Petroleum Publishing Company, 1972.

Howell, J. K. and Hogwood, ELECTRIFIED OIL PRODUCTION, Tulsa, Okla., Petroleum Publishing Company, 1962.

Interstate Oil Compact Commission (Engineering Committee), OIL AND GAS PRODUCTION, Norman, Oklahoma, University of Oklahoma Press, 1951.

Miller, Ernest C., PENNSYLVANIA HISTORY STUDIES No. 4, Gettysburg, Pennsylvania, Pennsylvania Historical Association, 1954.

Miller, Kenneth G., OIL & GAS FEDERAL INCOME TAXATION, Chicago, Ill., Commerce Clearing House, 1966.

Moore, Preston L., DRILLING PRACTICES MANUAL, Tulsa, Okla. Petroleum Publishing Company, 1974.

McCray, A. W., OIL WELL DRILLING TECHNOLOGY, Norman, Oklahoma, University of Oklahoma Press, 1958.

Porter, Hollis P., PETROLEUM DICTIONARY FOR OFFICE, FIELD AND FACTORY, 4th Edition, Houston, Texas, Gulf Publishing Company, 1948.

Sell, George, THE PETROLEUM INDUSTRY, London, Oxford University Press, 1963.

Stephens, M. M., and Spencer, O. F., PETROLEUM REFINING PROCESSES, University Park, Pennsylvania, Pennsylvania State University Press, 1958.

Uren, L. C., PETROLEUM PRODUCTION ENGINEERING (DEVELOPMENT), Fourth Edition, New York, N.Y., McGraw-Hill Book Company, 1956.

Wendland, Ray T., PETROCHEMICALS—THE NEW WORLD OF SYNTHETICS, New York, N.Y., Doubleday and Company, 1969.

Williams, Howard R., OIL AND GAS TERMS: ANNOTATED MANUAL OF LEGAL, ENGINEERING, AND TAX WORDS AND PHRASES, Cleveland, Ohio, Banks-Baldwin Law Publishing Company, 1957.

Williams and Myers, MANUAL OF OIL AND GAS TERMS, New York, N.Y., Matthew Bender and Company, 1964.

Williamson, H. E., et al, THE AMERICAN PETROLEUM INDUSTRY, Chicago, Ill., Northwestern University Press, 1963.

Zabo, Joseph, MODERN OIL-WELL PUMPING, Tulsa, Okla. Petroleum Publishing Company, 1962.